# ELECTRICAL REPAIRS MADE EASY

# ELECTRICAL REPAIRS

## MADE EASY

by Peter Jones

Butterick Publishing

**The Publisher believes that the material contained herein is accurate, but disclaims all responsibility in connection with its use or misuse**

**Library of Congress Cataloging in Publication Data**

*Jones, Peter, 1934-*
*Electrical repairs made easy.*

*Includes index.*
*1. Electric engineering — Amateurs' manuals.*
*2. Electric wiring, Interior — Amateurs' manuals.*
*3. Household appliances, Electric — Maintenance and repair — Amateurs' manuals. I. Title.*
*TK9901.J66 621.319'24'0288 79-25375*
*ISBN 0-88421-094-4*


Butterick Publishing
708 Third Avenue
New York, New York 10017
A Division of American Can Company


*Manufactured and printed in the United States of America.*
*Published simultaneously in the USA and Canada.*

# CONTENTS

CHAPTER 1 . . . . . . . . . . . page 7

How to Make Electrical Repairs

CHAPTER 2 . . . . . . . . . . . page 24

House Wiring Repairs

CHAPTER 3 . . . . . . . . . . . page 43

Preparing to Upgrade Your House Electrical System

CHAPTER 4 . . . . . . . . . . . page 49

Running New Circuits

CHAPTER 5 . . . . . . . . . . . page 75

Outdoor Wiring

CHAPTER 6 . . . . . . . . . . . page 86

Appliance Repairs

CHAPTER 7 . . . . . . . . . . . page 112

General Information

INDEX . . . . . . . . . . . page 117

# CHAPTER 1 • How to Make Electrical Repairs

ELECTRICITY IS EMINENTLY logical, more logical than any of the other manual arts. It leaves no room for error—if you do not twist the right wires together at the right places, nothing happens. Because you cannot see, hear, smell, touch , or taste electricity, it appears to be mysterious—but step into a puddle of water while holding the bare end of an electric wire and, without question, the electricity will make itself felt.

**Warning: Before you touch any wires, always shut off the circuit breaker or remove the fuse protecting the electrical line or lines you are repairing.**

## CIRCUIT INTERRUPTERS (Protective devices)

Circuit interrupters can be fuses, circuit breakers, or a master switch. Their purpose is to protect the electrical lines in your house from receiving so much current that the wires overheat and perhaps cause a fire. All protective devices are contained in metal boxes attached to a board nailed to a wall, usually in the basement, near the electrical service entrance. The service entrance is a large cable that branches off from the local electrical company's main cables, enters your house, passes through a kilowatt meter, and terminates at a main power panel. The main power panel is connected to a distribution box, which divides the electricity into several smaller, branch lines.

### *Main Power Panels*

The main power disconnecting switch in your home may be housed in its own metal box or may share a metal cabinet with the branch circuit protection devices (fuses, circuit breakers). However it is contained, it

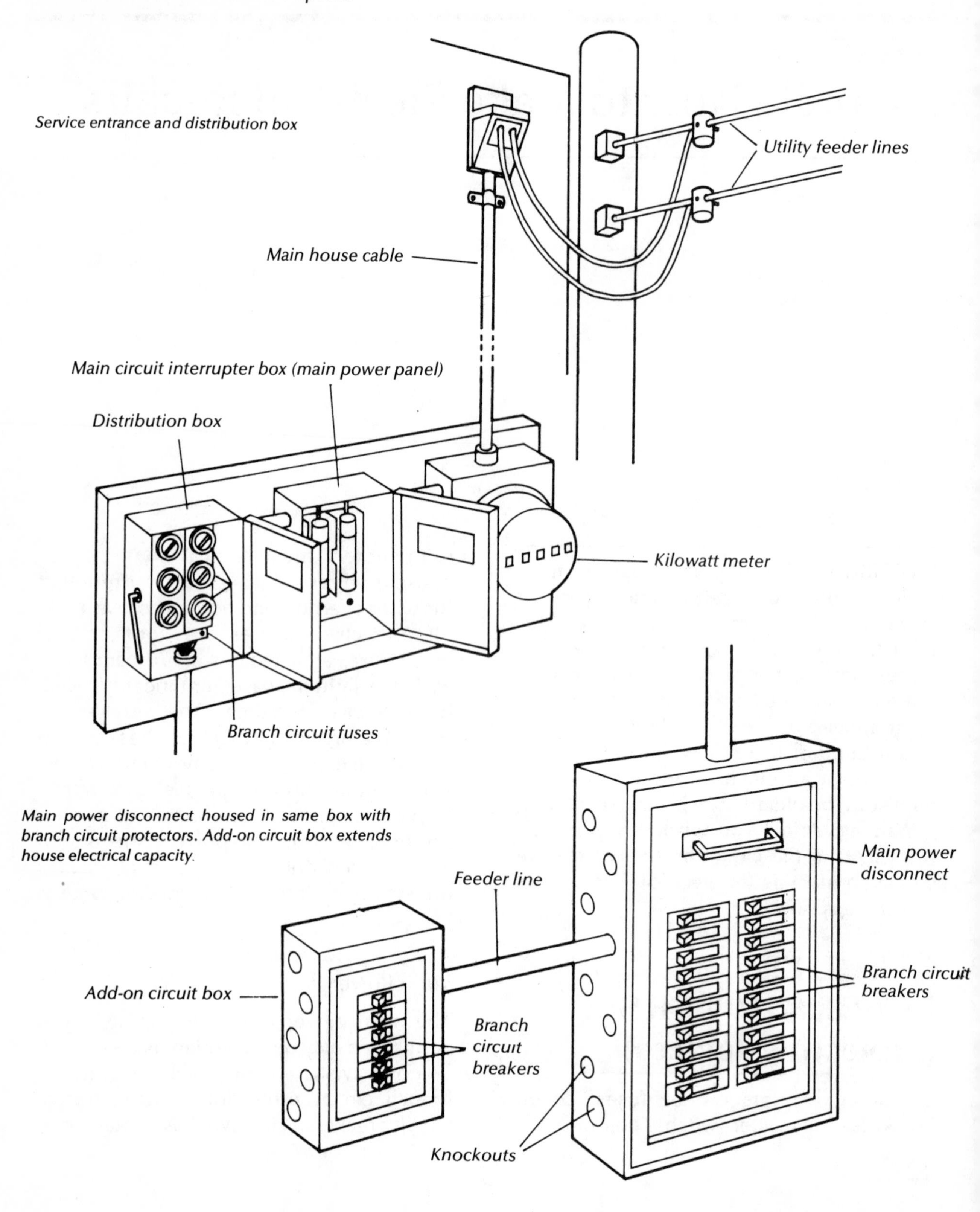


*Service entrance and distribution box*

*Main power disconnect housed in same box with branch circuit protectors. Add-on circuit box extends house electrical capacity.*

must be operated from outside the box, so you cannot even touch the disconnect until it has been deactivated. There are three types of main power disconnects: levers, drawer types, and circuit breakers.

*Lever-type main power disconnects*—Unless their electrical system has been upgraded, older homes have their main power disconnect in a separate cabinet, which contains fuses and has a lever attached to its side. Pulling the lever to its Off position unlocks the cabinet door, at the same time disconnecting all of the electricity entering the house so you can safely change one of the fuses in the box.

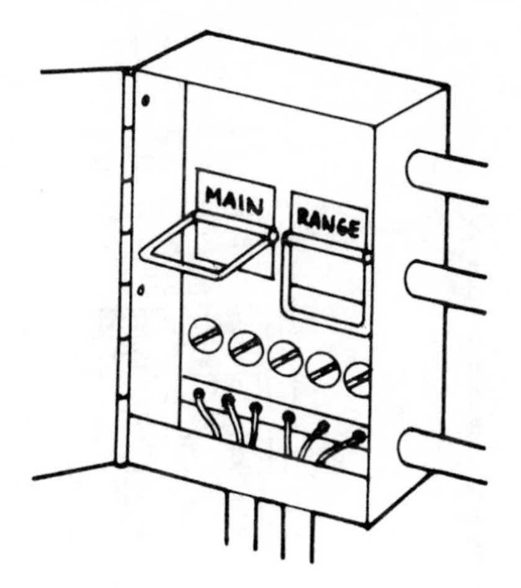


*Drawer-type main power disconnect with separate drawer for kitchen range*

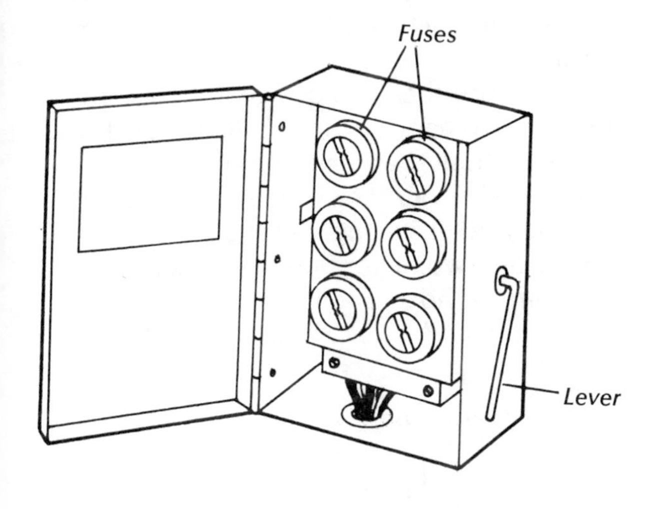


*Lever-type main power disconnect*

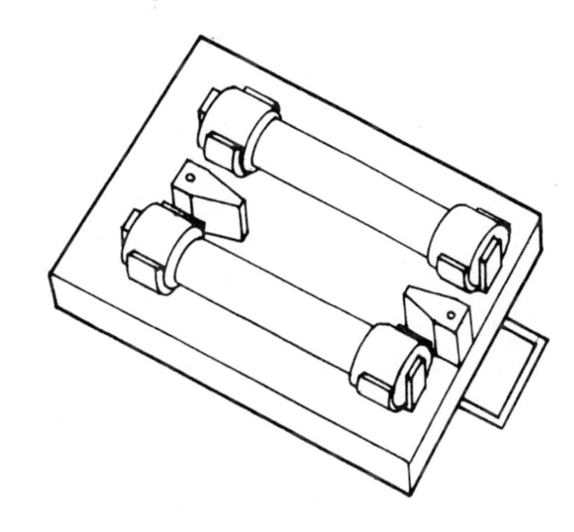

*Cartridge fuses on back of drawer*

*Drawer-type main power disconnects*—Main cartridge fuses (see cartridge fuses, page 13) are the most common main switch found today. They are mounted in a separate compartment inside the panel box. A handle on the face of the compartment permits you to remove the drawer, simultaneously pulling the fuses attached to the back of the handle out of their cavity, which automatically shuts off all of the power in the house.

Aside from the main switch, the panel box may contain separate drawers for such high-voltage appliances as your hot water heater, clothes dryer, and kitchen range. If you find more than one drawer switch in a power panel, each will be marked with embossed lettering on its face. To replace a cartridge fuse, pull the old fuse out of the clips that hold it and insert a new fuse. Then reinsert the drawer, right side up (according to the markings on its face).

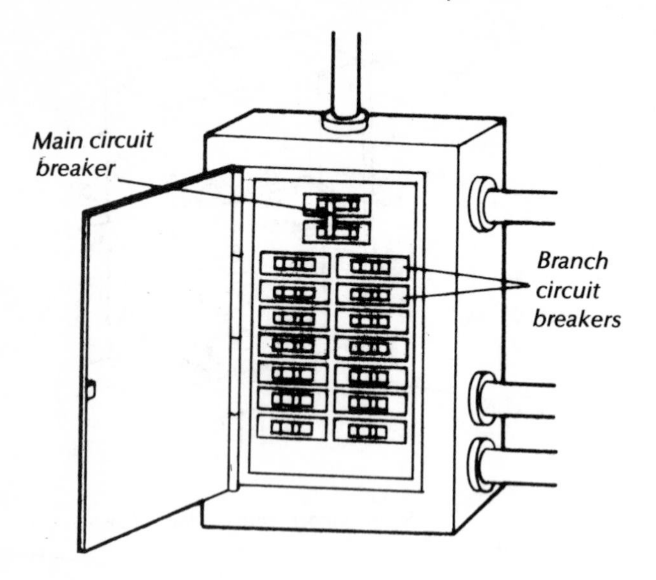


*Circuit breaker panel box*

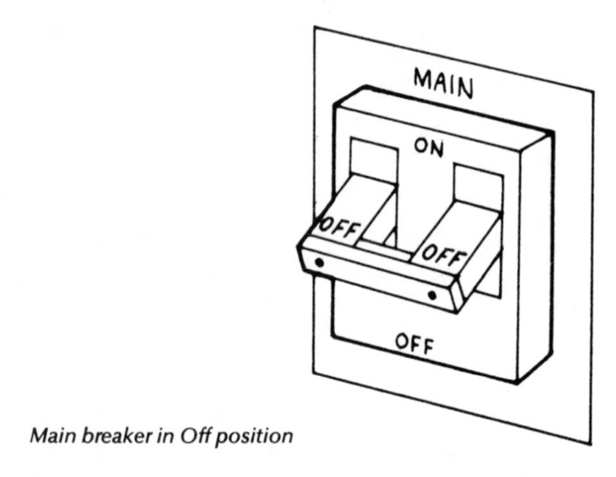


*Main breaker in Off position*

*Circuit breaker main disconnects*—Modern electrical installations incorporate circuit breaker panels, rather than a fuse box. When you open the panel door you are faced with tiers of toggle switches, each marked "Off" and "On." At either the top or bottom of the tiers is a separate, double switch (which is actually the handle of two circuit breakers linked together) labeled "Main."

To shut off all of the power entering the house, the double switch is pushed to its Off position. To restore the power, one type of circuit breaker is simply pushed to its On position. A second type is designed so that you must push the switch past its Off position to a Reset position and then push it back to On. With either type, the power is restored as soon as the toggle switch is in its On position.

## Fuses

Standard fuses contain a soft strip of metal through which current flows from the main power disconnect to the branch line, or cable, that the fuse protects. The metal strip is manufactured to handle a specific maximum amount of current; it will melt as soon as more than the designated amount of electricity passes through it. When the metal strip melts, it breaks the circuit so that no more electricity can enter the branch line, and the fuse is said to have blown.

There are two major reasons for a fuse to blow, and although it is easy enough to

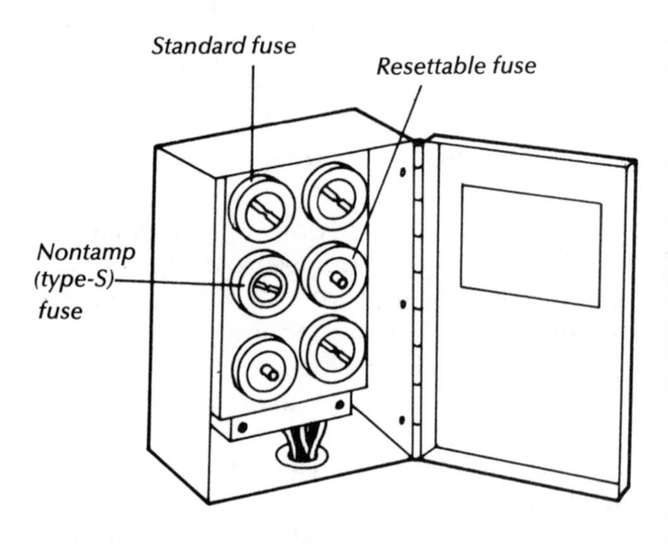


*A fuse box can have different types of fuses*

replace the fuse, that represents only half of the repair job. The other half is to locate the cause and correct it. A fuse may blow because:

- Too many appliances are plugged into the circuit and are drawing more current than the wires can handle. The solution to this problem is to remove some of the appliances. **It is dangerous to replace the blown fuse with a higher-rated fuse.**

- "Live" wires are touching each other, or part of a fixture, and are grounded (see page 27), causing a short circuit. This will often occur in the circuitry inside an appliance, rather than the house wiring system, so that the repair must be made in the appliance.

## Plug Fuses

These have a clear face, allowing you to see the metal strip inside, and a threaded base that looks like the bottom of a light bulb. They are screwed clockwise into threaded sockets in the panel board.

- If a plug fuse blows and you can see that the metal strip is broken, the cause is most likely an overloaded circuit.
- If a plug fuse blows and its window turns black, so that you cannot see the metal strip, the cause is probably a short circuit.

**Be sure to replace a blown fuse with a fuse of the same amperage rating.** The amperage is clearly marked on the face of each fuse and will be 15, 20, 25, or 30 amps. Because it is so easy to replace a plug fuse with a fuse of a higher rating, the National Electrical Code (see page 49) has prohibited the installation of plug fuses in any new construction.

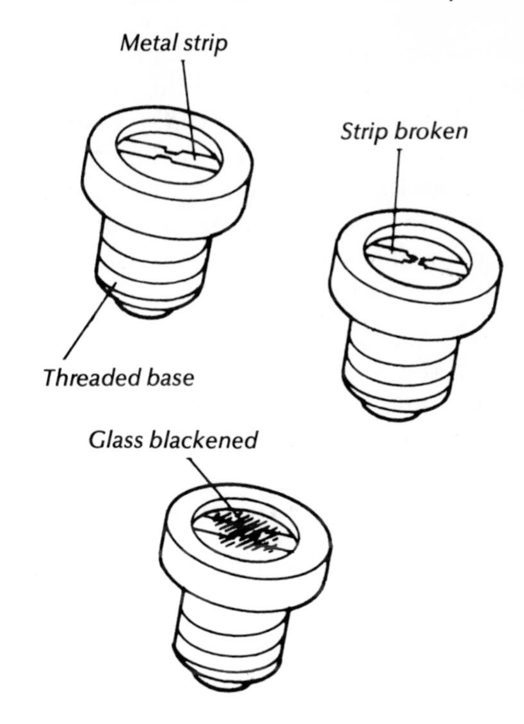


Plug fuse

*Nontamperable (type-S) fuses*—Nontamp fuses consist of a screw-in fuse and an adapter, which is a threaded base that screws into any plug fuse socket. Once installed, the adapter cannot be removed, and it will accept only a type-S fuse of an identical rating. Thus, a 15-amp adapter will accept only a 15-amp type-S fuse; a 20-amp adapter will take only a 20-amp type-S fuse. The entire purpose of type-S fuses is to prevent anyone from overfusing the circuits they protect.

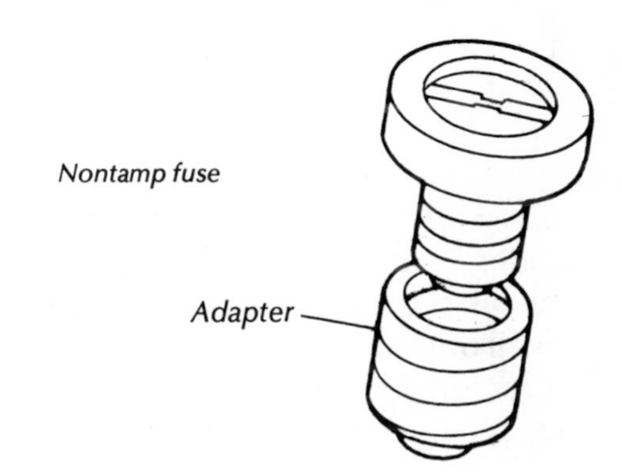

When inserting a type-S fuse, be certain you tighten it as far as possible. The spring metal strip under the head of the fuse must be flattened for the fuse to make complete contact and allow current to flow through it. Even after you have tightened the type-S fuse as much as you think is necessary, try to turn it some more.

*Time-delay (Fusetron) fuses*—Another type of plug fuse, these have a spring-loaded metal strip that allows them to accept a momentary power overload without blowing. Motor-driven appliances, such as stationary power tools, refrigerators, and air conditioners, draw extra power (surge) for a few seconds when they start up. This initial power requirement may be more than the rated capacity of the circuit, and would cause ordinary fuses to blow repeatedly even though the circuit was in no danger of being overloaded beyond the limits of safety.

You can safely replace any standard plug fuse with a time-delay fuse; time-delay fuses will blow just as quickly as standard plug fuses if a short circuit occurs, or if the line is overloaded for an extended period of time.

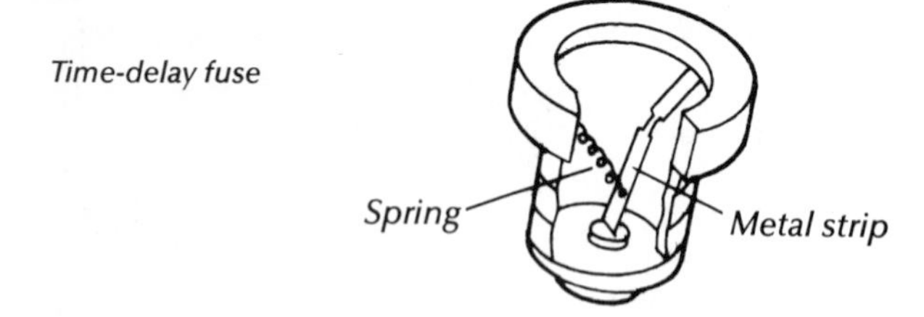


*Time-delay fuse*

*Screw-in breaker (resettable) fuses*—Any standard plug fuse can be replaced with a screw-in breaker fuse. The breaker fuse has a small button in the center of its face that is depressed to complete the electrical circuit. The moment a short circuit or overload occurs, the button pops out. To restore power, you merely have to push the button in again. Although screw-in breaker fuses cost as much as $3, they never have to be replaced, and they are as convenient to use as a circuit breaker. They are particularly useful on circuits serving motor-driven appliances, such as might be found in a workshop.

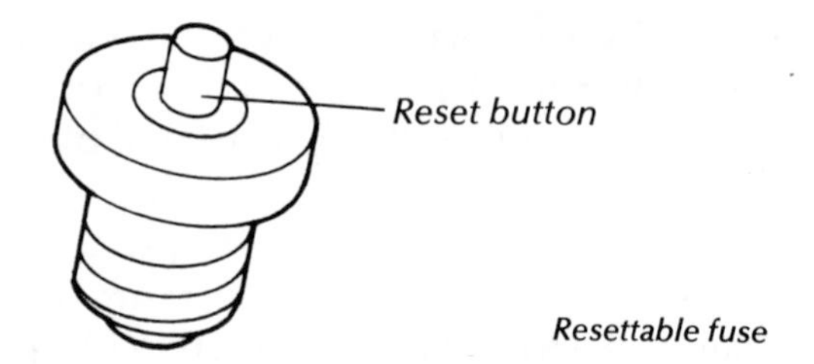


*Resettable fuse*

## *Changing a blown plug fuse*

All plug fuses are replaced the same way:

1. **Shut off the main power.**
2. Open the fuse box and examine the fuses to determine which one has blown.
3. Refer to the label on the inside of the fuse box cover to locate the switches and outlets on the circuit controlled by the blown fuse. Turn off all switches and unplug all lamps and appliances on the circuit. (Otherwise, the new fuse might blow if all of the current-drawers came on simultaneously.)
4. Unscrew the blown fuse by rotating it counterclockwise.
5. Screw a replacement fuse into the socket. Be sure the new fuse has the same ampere rating as the one you are replacing.
6. Close the fuse box.
7. Check out the circuit (see page 14).

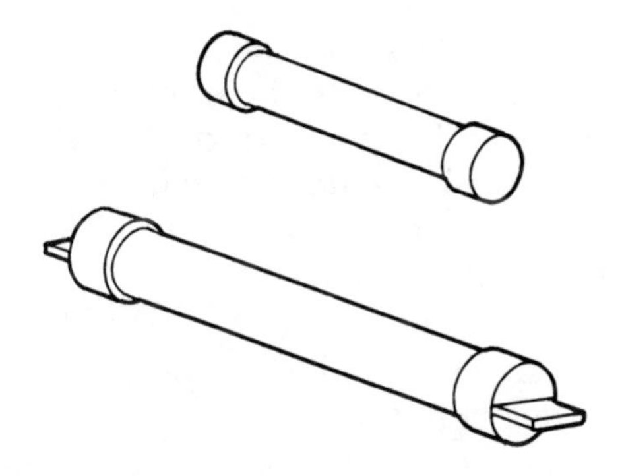

*Standard and knife-edge-contact cartridge fuses*

### Cartridge Fuses

There are two types of cartridge fuses. The cylindrical version is made to handle up to 60 amps. The second type has a knife-edge contact at each end and is designed to handle 60 amps or more. Cartridge fuses are usually found only in the main power line or in high-voltage appliance circuits and are installed in drawers. However, some older homes have cartridge fuses on their individual branch circuits.

**The meter-side end of a cartridge fuse is always live, even when the fuse has blown. Do not touch it.**

#### *Changing a cartridge fuse*

You can purchase fuse pullers for removing cartridge fuses safely. If you do not have a fuse puller:

1. **Pull the main disconnect switch to its Off position.**
2. Open the fuse box.
3. You have no way to tell whether a cartridge fuse has blown unless you know which circuits it controls—or you have a hot-line tester (see page 50). To test a cartridge fuse with a hot-line tester, touch one probe to the white, return wire at the bottom of the panel box and the other probe to the black wire on the incoming side of the fuse. If the light goes on, the fuse is receiving power. Next, touch one probe of the tester to the white wire and the other probe to the exit (black-wire) end of the fuse. If the light goes on, power is leaving the fuse. If it does not go on, the fuse has blown.
4. Turn off all switches and unplug all lamps and appliances on the circuit.
5. Use a wooden stick to pry the fuse from its spring-clamp holder. **Let the fuse fall to the floor—do not touch it.**
6. The replacement fuse must have the same amperage rating on it as the old fuse. Push the replacement fuse into the holder clamp.
7. Close the fuse box cover.
8. Move the main switch to its On position.
9. Check out the circuit (see page 14).

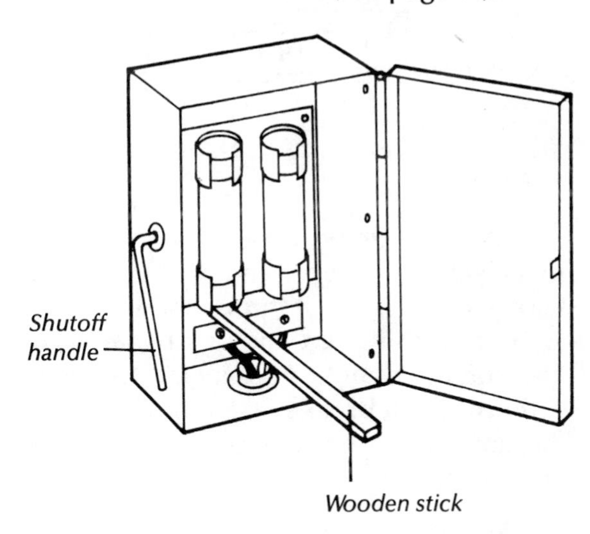


*Pry the fuse out with a wooden stick*

## Circuit Breakers

Each circuit breaker is, in essence, a toggle switch similar to any wall switch. It is

*Circuit breaker panel box. Circuit breakers can be toggle switches or push buttons*

wired to the circuit breaker panel box so that electricity enters its hot side via a black wire and returns to it through a white wire connected to the grounding busbar at the base of the panel. Trouble rarely occurs within a circuit breaker itself, but if it does, the unit should be replaced by a licensed electrician.

### *Resetting a circuit breaker*

1. Open the cover of the circuit breaker panel box and examine all of the circuit breakers, looking for any handles that are in the center, or Off, position, or for a push button that extends out farther than the others.
2. Check the label on the inside of the panel box cover to determine which switches and outlets are on the circuit controlled by the tripped circuit breaker. Turn off all switches and unplug all lamps and appliances on the circuit.
3. Push the circuit breaker handle to Reset or full Off, then to On; or push the circuit breaker button.
4. If the circuit breaker does not stay on, wait 60 seconds and turn it on again. If it still does not stay on, there is probably a short circuit somewhere in the house circuit (since all of the appliances were taken off the circuit), and each of the switches and outlets will have to be painstakingly checked for wires touching each other or the electrical boxes that contain them (see page 27).

   If the breaker remains on, close the panel cover.
5. Check out the entire circuit.

## Circuit Check

Every time a fuse blows or a circuit breaker trips, the first step is to restore power to the house circuit. The second step is to locate and correct the cause of the power interruption. In checking out a circuit you are trying to determine whether there is a short circuit in a switch, outlet, or appliance, or whether there are too many appliances on the circuit to begin with. After you have replaced or reset the circuit interrupter, check out the circuit.

1. One at a time, turn each switch on, then off.
2. One at a time, plug in each appliance and turn it on, then off.

   If at any time the circuit interrupter again blows or trips, the switch or appliance you have just activated is defective and should be repaired.

3. When you have individually checked each of the devices on the circuit, turn them all on at the same time. If the circuit interrupter blows or trips, too many appliances are overloading the circuit. Take two or three of them off the circuit. If you can't plug them into another circuit (in the kitchen, for example), consider adding another branch circuit to service the extra appliances. (See Chapter 3 for how to install branch circuits.)

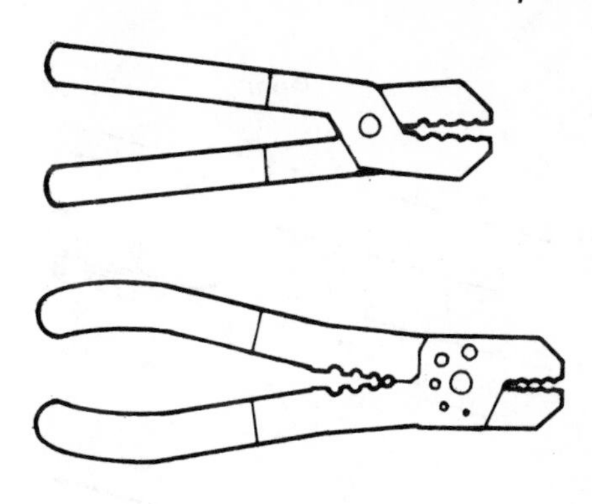

*Wire stripper (top) and multipurpose tool*

# CONNECTING WIRES

In nearly every instance, electrical repairs consist of connecting wires either to terminals or to other wires. There are several acceptable methods for making safe wire connections.

## Preparing Wire for Connection

Electrical wire is always fully insulated with a coating of rubber or plastic, which must be stripped (removed) before the wire can be properly connected to anything. The amount of insulation to be stripped from the end of the wire depends on the kind of connection you are going to make with it. In most cases, between ½″ and 3″ of bared wire is sufficient.

The most efficient as well as least damaging method of stripping insulation from any wire is to use wire strippers, or a multipurpose tool, which have a series of sharp openings in their scissors blades to allow stripping any diameter of wire. They will not cut into, thereby weakening, the wire. They are also quicker and easier to use than a knife.

### *Using wire strippers*

1. Open the strippers and insert the wire end in the proper diameter hole.
2. Close the strippers around the wire so that the tool cuts through the insulation but not the wire.
3. Pull the insulation from the end of the wire.

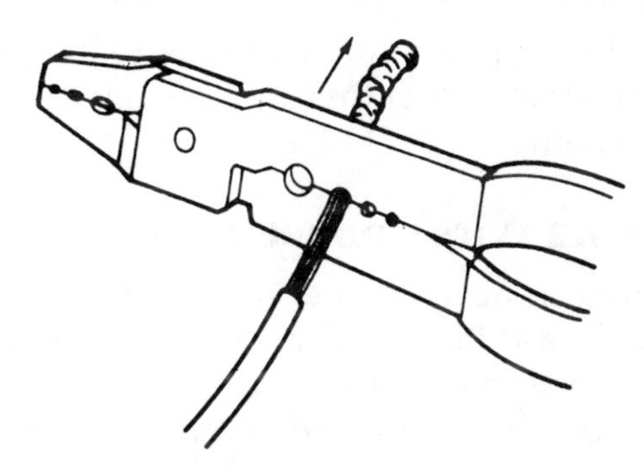

*Stripping insulation from stranded wire*

### *Stripping wires with a knife*

If you are stripping wire with a knife, be careful not to nick the wire and weaken it.

1. Cut the insulation completely around the wire. Do not cut straight into the insulation, but slant your knife blade at a 60° angle toward the end of the wire. Cutting at an angle does not guarantee that you will avoid nicking the wire, but it reduces the chances considerably.
2. Pull the insulation off the wire, using your fingers or a pair of pliers.

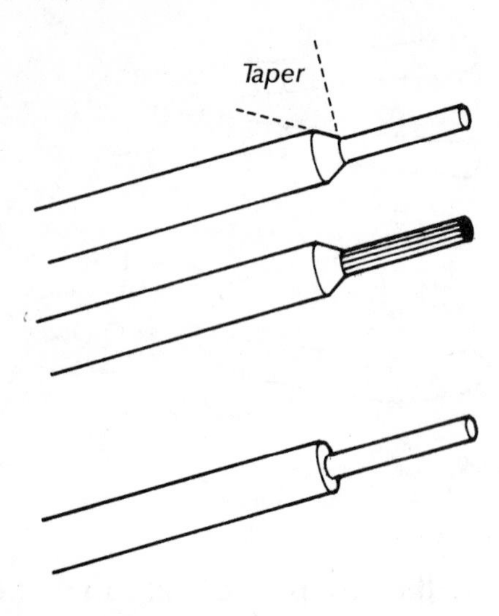


Cut insulation on a slant, not perpendicular to wire

3. Scrape the wire with the back of the knife blade to clean it of all insulation.
4. If the wire is stranded (see page 52), twist the strands together in a clockwise direction.

## Making Wire Connections

Wire can be spliced to other wire of the same or a different diameter; twisted together with another wire and held with a wire connector; looped around a screw terminal; or joined with a crimp-type connector. Connections can also be twisted together and taped or soldered, although neither procedure is recommended for house wiring. When you are working with the wires in your house electrical system, most often you will attach wire to a screw terminal or splice wires together with a wire nut.

### *Connecting to screw terminals*

1. Remove only enough insulation to make a loop of bare wire around the screw terminal. This should be about ½".
2. Using long-nosed pliers, bend the bared wire into a loop. If the wire is stranded (see page 52), the strands should first be tightly twisted together, and can even be soldered.

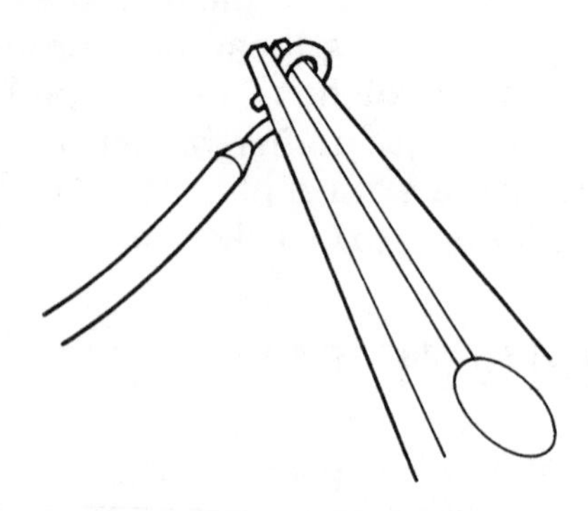

Bend the bared wire into a loop

3. Loosen the screw terminal with a screwdriver, but do not remove it from its hole.
4. Hook the wire over the top of the screw so that when you tighten the screw the wire will be drawn around it, not pushed away from the screw threads.

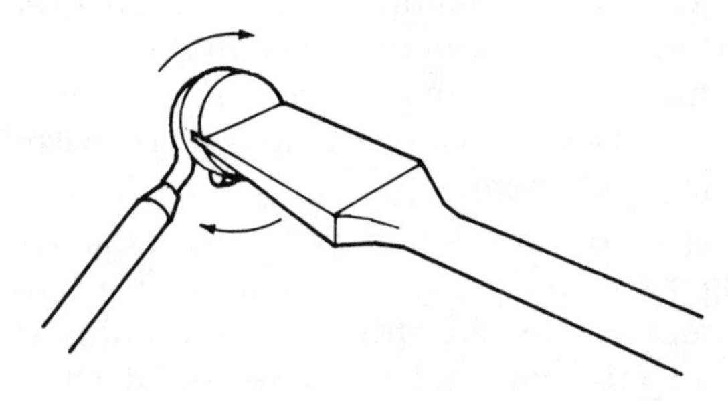

The loop is attached clockwise, as the screw is tightened

5. Close the loop around the screw with pliers.
6. Tighten the screw by turning it clockwise. Almost no bared wire should extend beyond the screw head. If it does, you have stripped away too much

insulation and the bared wire should be shortened.

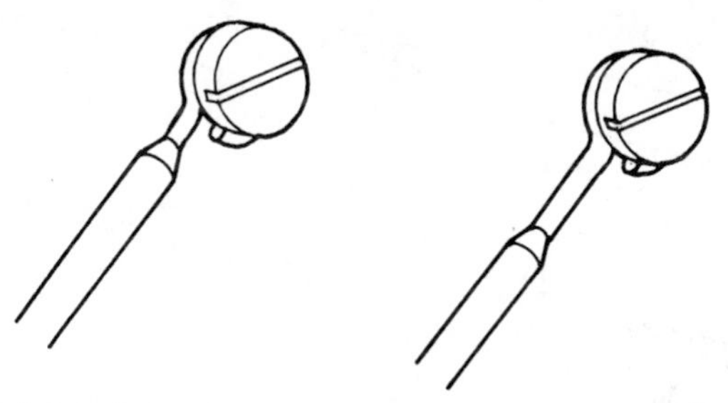

*Little bared wire should extend beyond the screw head (left). If too much wire shows (right), take the loop off and recut it.*

## Using Wire Nuts

Wire nuts are designed to hold two or more bared wires together when there will be no strain on the wires; in a house electrical system, all wire splices are contained in electrical boxes, where they are protected from any strain by clamps (see pages 63–64). There are two types of wire nuts. One is an uninsulated spring, which must be wrapped with electrician's tape to complete the connection. The other, more common type is a spring housed in a plastic cap. With either type, you must use a connector that matches the size of the wires being spliced; they can be purchased in a variety of diameters and are capable of holding more than two wires together.

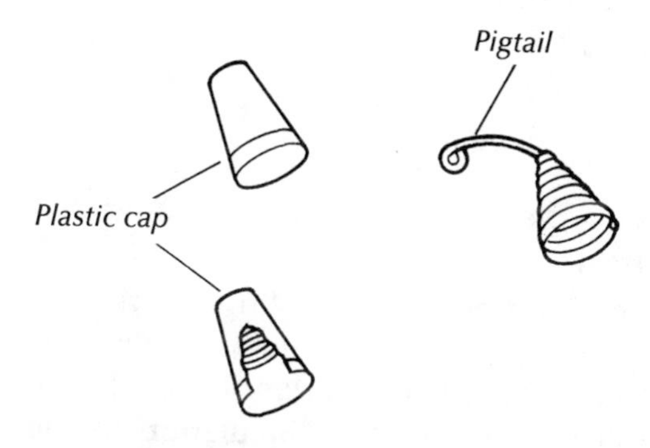


*Insulated (left) and spring-type wire nuts*

### Connecting with an insulated wire nut

1. Strip approximately ¾″ of insulation from the ends of the wires to be spliced. If one of the wires is a smaller diameter than the other(s), strip about ¼″ more insulation from it.
2. Twist the bared wires together. You may need to use pliers in order to make the twist tight and firm.

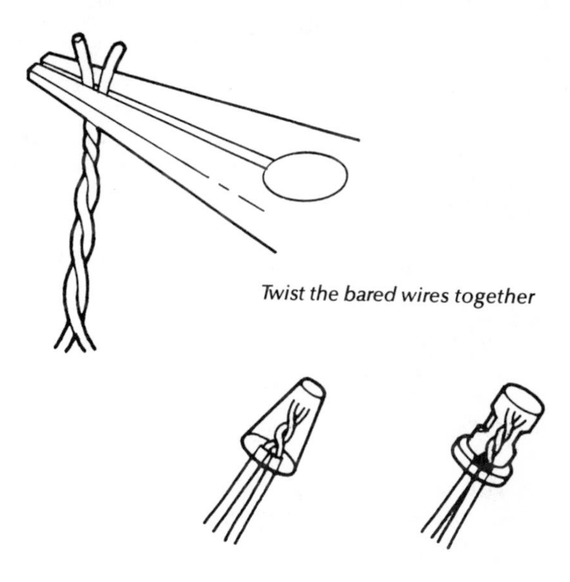

*Twist the bared wires together*

*The wire nut must cover all of the bared wire*

3. Push the wire nut down over the bared wires and rotate it clockwise. It must cover *all* of the bared wire.
4. The wire nut may be wrapped in electrician's tape if you so desire, but taping is usually not necessary unless there is the possibility that the wires might be pulled apart.

### Connecting with an uninsulated wire nut

1. Strip about ¾″ of insulation from the wires to be connected. If one of the

wires is a smaller diameter, strip about ¼″ more insulation from it.

2. Hold the bared wires next to each other and twist the wire nut clockwise over them.

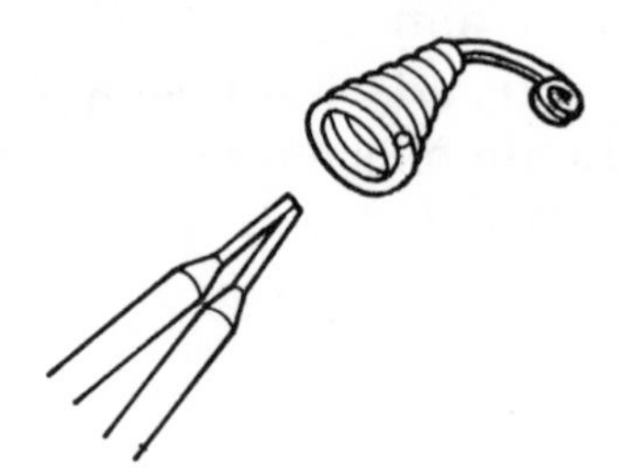

*It is not necessary to twist the wire ends together before inserting in an uninsulated wire nut. Clip off the pigtail after the wires are connected*

3. Break off the handle attached to the wire nut.
4. Wrap the wires and wire nut with electrician's tape.

## Splicing Wires

There are three criteria to bear in mind when you are splicing wires:

- The connection must be as strong as if the wire were unbroken.
- The connection must be as electrically sound as if the wire had never been broken.
- The connection must be insulated by at least the same thickness of covering as the insulation on the rest of the wires.

### Splicing solid wires

1. Strip approximately 3″ of insulation from each wire end.
2. Using a knife, slice a long taper (about 20°) in the insulation to facilitate making a smooth covering of electrician's tape around the splice.

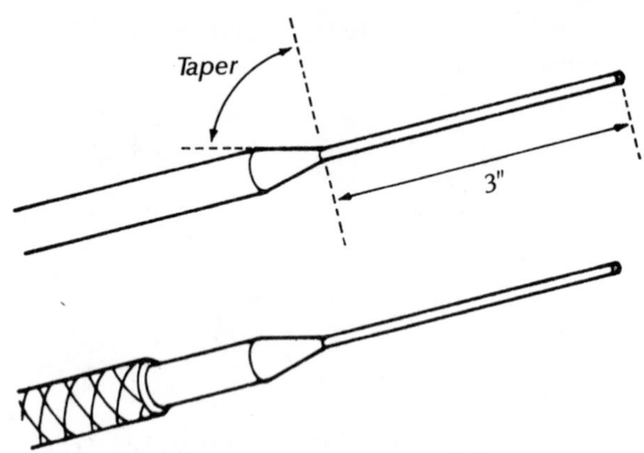


*Slice a long taper in the insulation*

3. At about 1″ from the end of the insulation, cross the bared wires and bend their ends back toward the insulation.

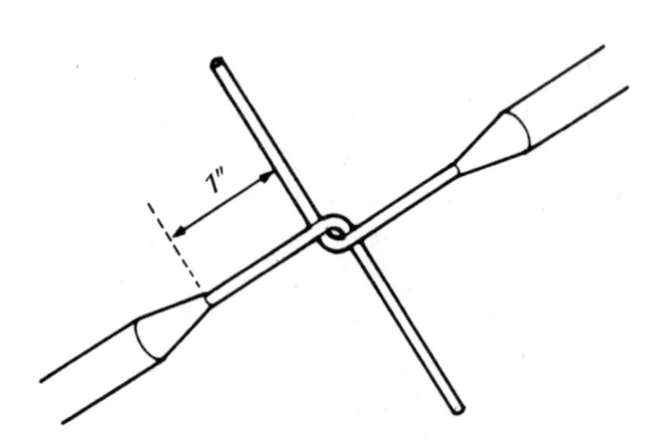


*Cross the bared wire ends and bend them back*

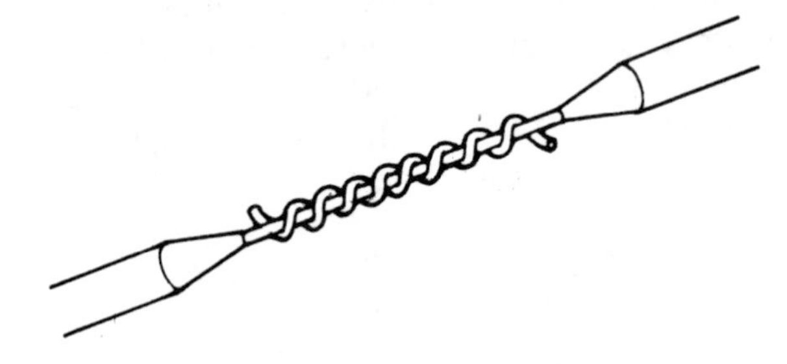

*Twist the ends around the wires*

4. Twist the two free ends around the wires.
5. Beginning on the insulation at one end of the splice, wind electrician's tape around the splice, overlapping each turn until you overlap the insulation at the other end of the connection. Then wrap back over the splice. Continue wrapping

back and forth around the splice until the thickness of the tape equals that of the insulation around the wires.

*Splicing stranded wires*

1. Strip about 3″ of insulation from the ends of the wires to be spliced and taper the insulation at a 20° angle.
2. Spread the strands in both wires evenly in all directions.
3. Push the splayed wires together so that the strands all cross each other.
4. Wrap one strand of one of the wires tightly around all of the strands in the second wire.
5. Wrap one strand of the second wire around all of the strands in the first wire.
6. Continue alternately wrapping strands around the wires until all of the strands are twisted around the splice.
7. Wrap the splice with electrician's tape until the thickness of the tape equals that of the insulation around the wires.

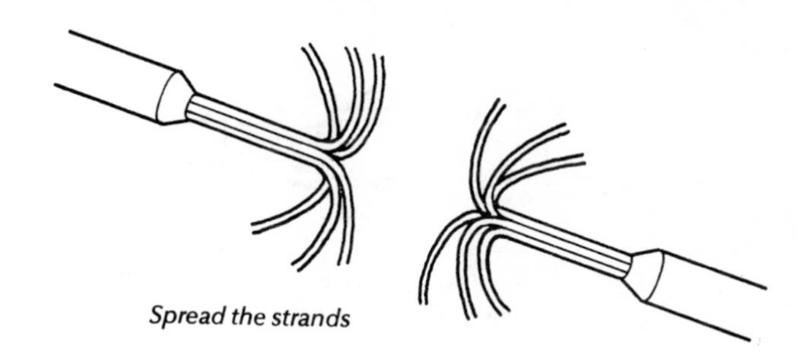

*Spread the strands*

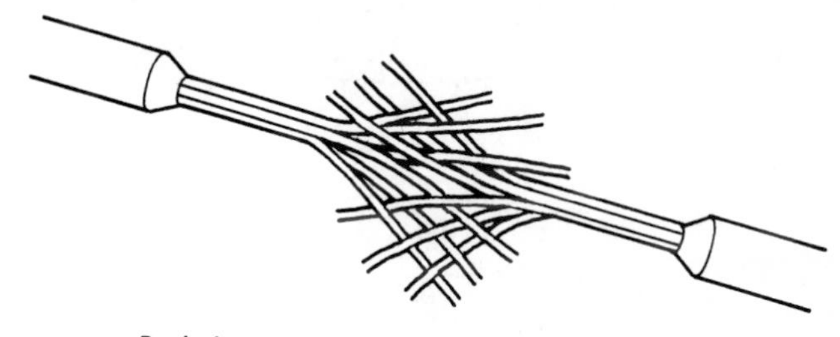

*Push the splayed strands together*

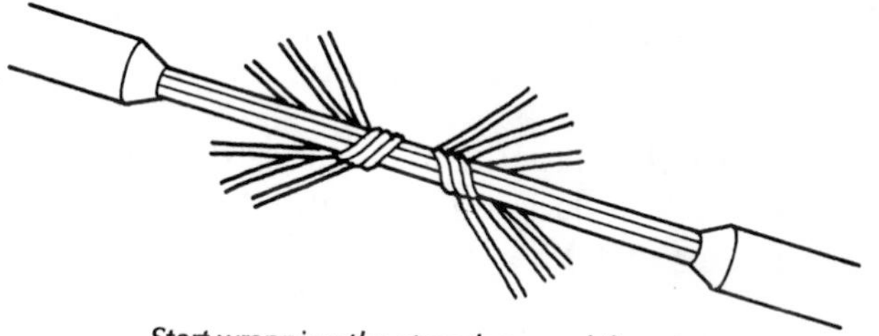

*Start wrapping the strands around the wires*

*Finish wrapping all of the strands*

*Splicing cord wires*

Splicing a two-wire cord is no different from splicing any other stranded or unstranded wires, except that when you have spliced both wires in the cord you have effectively made two splices of bared wire that are next to each other but *must not touch*. The splices must be completely insulated from each other. Consequently, the most important part of splicing a power or extension cord is how you wrap it with electrician's tape.

1. Strip approximately 1″ of insulation from each of the wires to be spliced. With most appliance and extension cords this means you will be baring four wires. (In a three-wire cord, the third wire is for grounding purposes.) The cord wires are individually insulated and then wrapped in a common sheath; therefore, you will wrap your splices individually and then together.
2. Twist the wires together, making two (or three) separate splices.
3. Wrap each splice with electrician's tape, overlapping the tape around the insula-

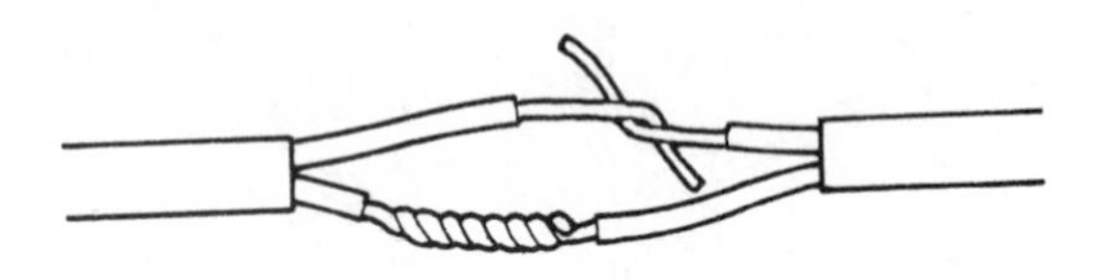

*Twist cord wires together in separate splices. The splices should not be exactly opposite each other*

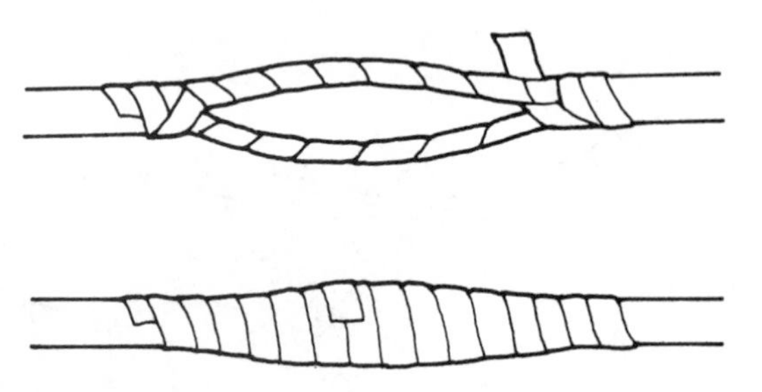

*Tape each splice separately (top), then tape the splices together*

tion at both ends of the splice. Be very careful to cover all of the bare wire.

4. Wrap the taped splices together with tape, overlapping it until the splice equals the thickness of the cord.

## Tapping Wires

You are tapping a wire when you connect one wire to an unbroken second wire. You should not make a tap if there is a chance that the wires will be under any strain. Thus, you can safely tap the wires inside an electrical box—although in most situations it is easier to cut the wires and reconnect them with wire nuts.

### Tapping solid wires

1. Strip 1″ of insulation from the wire to be tapped, slicing around the wire with a knife at two points 1″ apart, then slicing lengthwise between the cuts. You can then pull the section of insulation away from the wire.
2. Strip 1½″ of insulation from the end of the tapping wire.
3. Hook the tapping wire over the bared unbroken wire, approximately ¼″ from the insulation, and wrap it tightly around the unbroken wire.
4. Begin wrapping electrician's tape at one end of the connection. Wrap the tape around all of the bared wire, twisting it if necessary in order to cover the tapping wire.

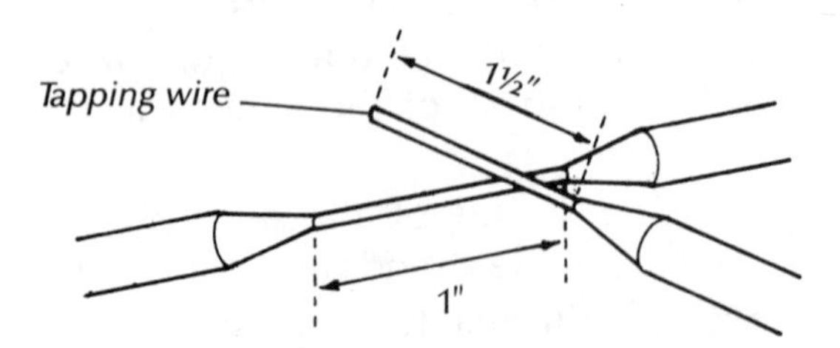


*Cross the tapping wire over the wire to be tapped*

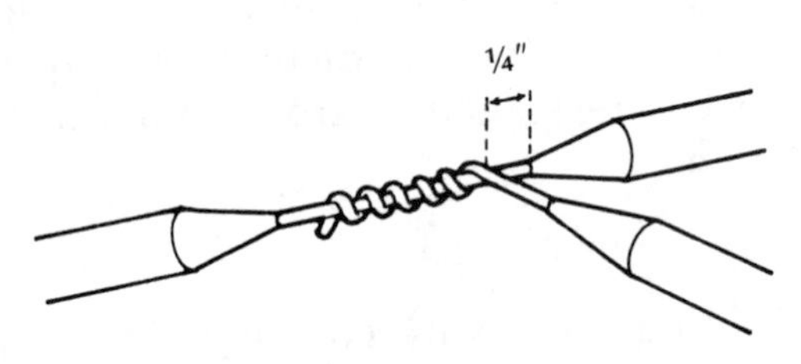


*Wrap the tapping wire around the wire to be tapped*

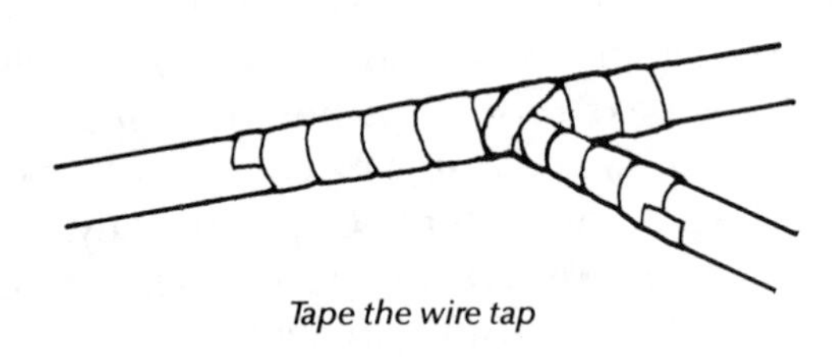

*Tape the wire tap*

### Tapping stranded wires

1. Strip 2″ of insulation from the wire to be tapped.
2. Strip 3″ of insulation from the end of the tapping wire.

3. Spread the strands in the unbroken wire, dividing them in half.

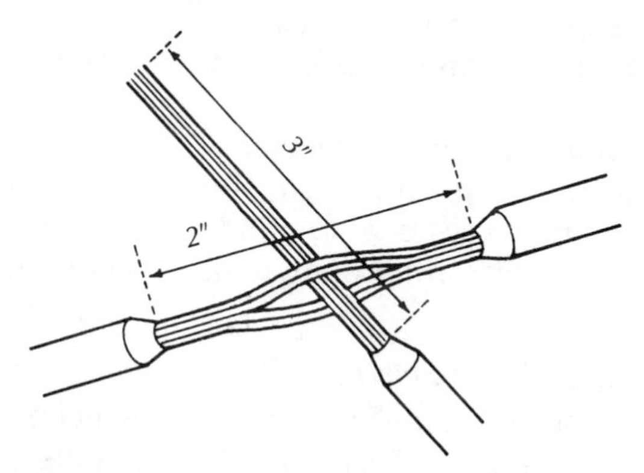


*Divide the strands of the wire to be tapped and insert the tapping wire*

4. Push the tapping wire between the divided strands of the unbroken wire.

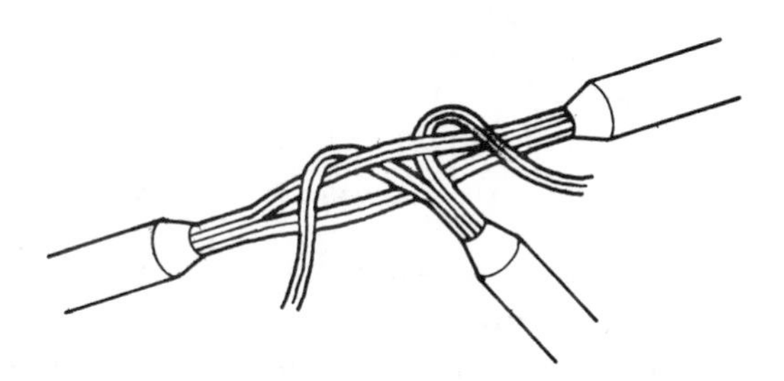

*Bend the strands of the tapping wire*

5. Divide the strands of the tapping wire in half and bend each group around the strands of the unbroken wire, in opposite directions.

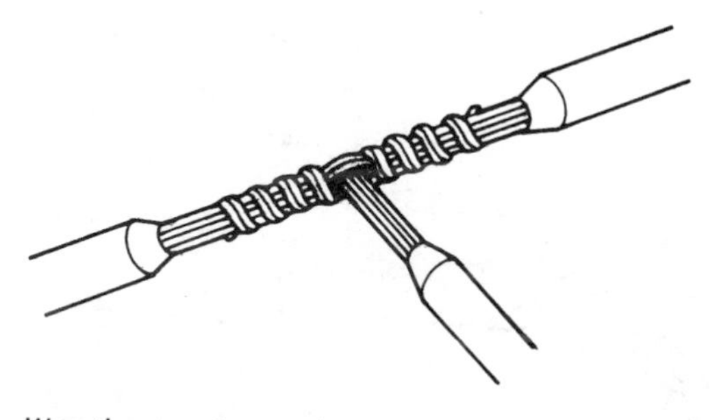

*Wrap the strands completely around the unbroken wire*

6. Wrap the two groups of strands in the tapping wire around the strands of the unbroken wire. Both groups should be wound in a counterclockwise direction.
7. Wrap the connection with electrician's tape.

## Crimp-Type Connectors

You can make your life as an electrician considerably easier by connecting wires with crimp-type splices and terminals. This is particularly true when you are working with smaller wires than are found in house electrical systems—for example, in appliances. Note that you cannot attach the connector to unsoldered stranded wire.

To install a crimp-type connector, you must have a crimping tool or a multipurpose tool with which to pinch the connector onto the bared wire end(s). The procedure for attaching both crimp-type connectors is the same.

1. Strip ½" of insulation from the end of the wire.
2. Slide the splice or terminal over the bared wire. The connector must be the same gauge (interior diameter) as the wire you are installing it on.

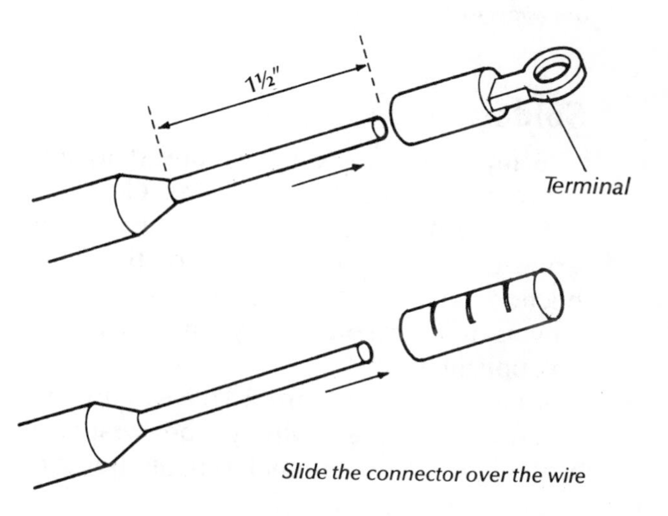


*Slide the connector over the wire*

3. Select the proper slot for the connector in the crimping tool, place the tool around the connector, and squeeze its handles together as hard as you can.
4. Yank on the terminal or wire to be certain the connector is secure on the wire. If the connection comes apart, repeat the procedure with a new connector.

Crimp-type terminals have either a hooked or a doughnut-shaped flange that fits around a screw terminal. The splices must be crimped twice after the wires are inserted in it, once at each end.

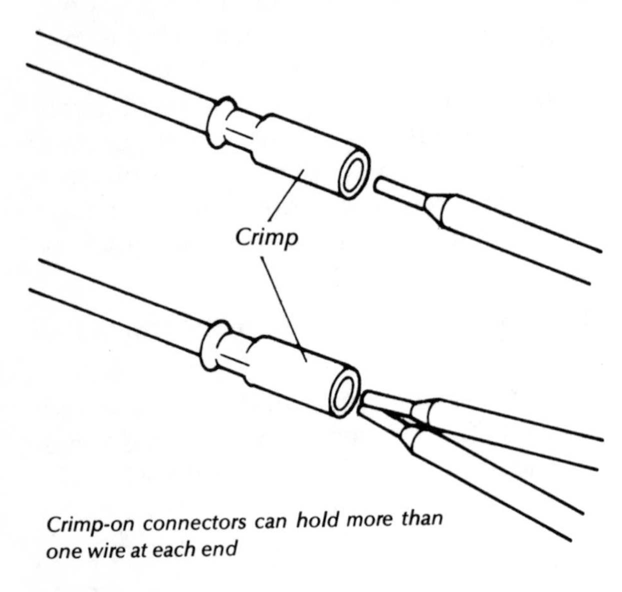


*Crimp-on connectors can hold more than one wire at each end*

## Soldering

Connections are often soldered to improve their strength as well as electrical contact. However, the National Electrical Code prohibits soldering any of the connections made in a house electric system, so you need to solder only when working on appliances.

A tapped connection, a splice, and the end of a stranded wire to be wrapped around a screw terminal can all be soldered. Solder is probably the best way of keeping stranded wire together so that it does not spread out in all directions when squeezed under the head of a terminal screw.

Soldering is not particularly difficult, but it does take practice, and there are three things to always bear in mind: cleanliness, enough heat, and proper application of the solder.

1. Strip the wires to be spliced.
2. Clean the bared wires of all insulation with steel wool, sandpaper, or an emery cloth.

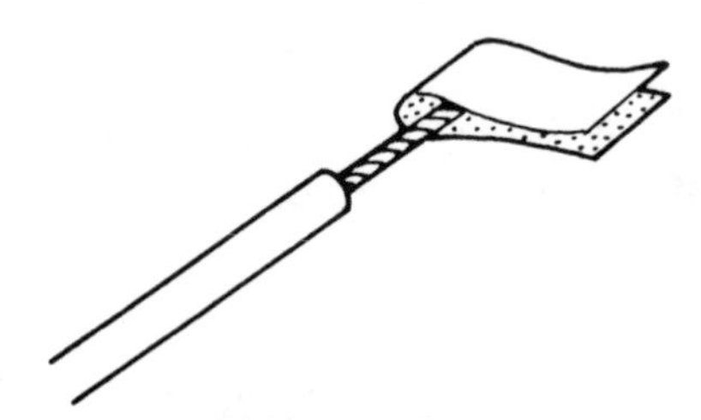

*Clean bared wire with a fine abrasive*

3. Twist the wires together with pliers. Do not use your fingers—any trace of skin oil on the metal might prevent the solder from adhering properly.
4. The tip of the soldering iron must be absolutely clean. If it is dirty, abrade it with a fine-grade metal file, emery cloth, or sandpaper until it is shiny.

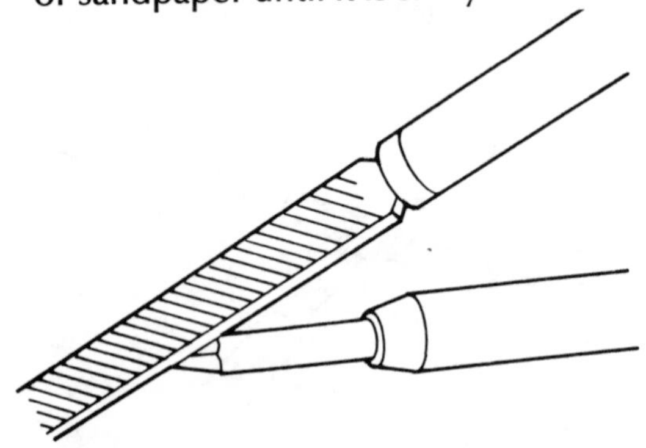

*Cleaning the soldering iron tip with a file*

5. Plug in the soldering iron and allow it to reach its operating temperature. With a soldering iron, this will take three to five minutes. Soldering guns require less than a minute.
6. When the soldering iron has reached its operating temperature, touch its tip to the end of a roll of rosin-flux solder. (Do *not* use acid-core solder, which is used to join metals structurally, not to splice electrical connections.) The solder will smoke for a few moments as the flux is "cooked" out. When the smoke stops, wipe the tip of the iron with a damp rag or sponge. This is called "tinning the tip," and it is an absolutely necessary step if you want the solder to adhere properly.

*Touch the heated tip to the solder*

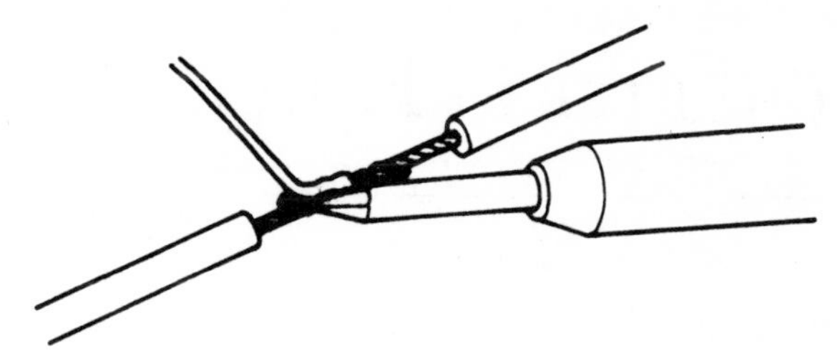

*The heat of the wires melts the solder*

7. Hold the tip of the soldering iron against the wires in the splice until they have heated enough to melt the solder.
8. Still holding the tip of the iron to the wires, touch the end of the solder to the splice. The solder will melt and run into the splice, completely covering even its underside. As soon as the splice is covered, remove both the solder and the soldering iron.
9. Allow the splice to cool, then gently pull at the wires to be certain the connection is secure. If the wires come apart, reheat them and apply more solder.
10. Wrap the connection with electrician's tape.

# CHAPTER 2 • House Wiring Repairs

THE KINDS OF REPAIRS needed in your house electrical system are basic in nature, and very few in number. Occasionally an outlet (receptacle) fails to function and should be replaced; switches sometimes stop working, usually because of a mechanical failure, and also need replacement; overhead fixtures (ceiling lights) are often replaced, usually for decorative purposes. Although lamps and their cords are not actually part of the house system, their repair can be included among home wiring repairs.

## SWITCHES AND OUTLETS

About a dozen switches and several types of outlets are used in home wiring systems, each housed in its own electrical box. How the electrical boxes are installed on their branch circuits is discussed in Chapter 4; so far as repairing or replacing a switch or outlet is concerned, it is necessary to know only that each electrical box has at least one electrical cable entering it. (A cable has two or more individually insulated wires, and sometimes a bare grounding wire.) The cable is clamped to the box so that it cannot move, and wires inside the cable extend about 8″ beyond the cable sheathing and into the box. The last ½″ of the wires is stripped of insulation and looped around screw terminals on the sides of the switch or outlet.

Both the wires and the screw terminals are color-coded. When you connect a wire to a screw terminal, the color of the insulation around the wire should match the color of the screw terminal.

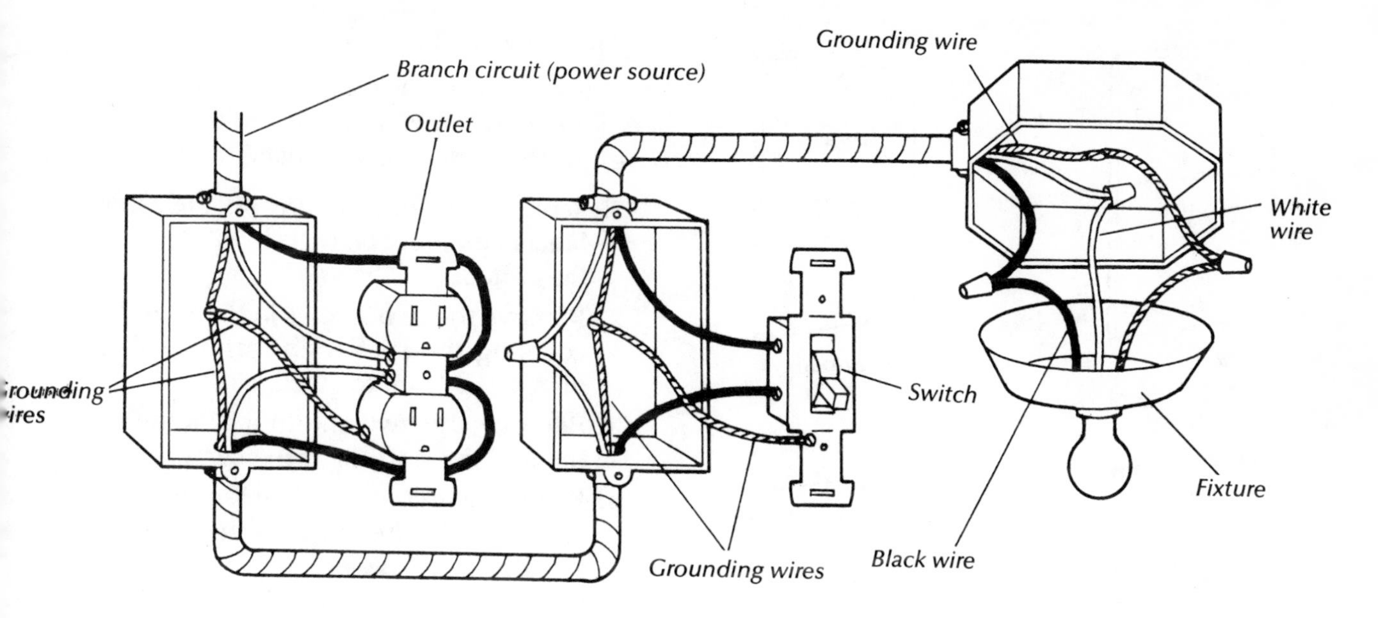


*The electrical equipment connected directly to a house wiring system is almost exclusively switches, outlets, and light fixtures*

| WIRE COLOR | SCREW COLOR |
|---|---|
| Green | Green (or any dark color) |
| Red | Brass |
| White | Silver (chrome) |
| Black | Brass |
| Bare wire (any color) | Connected to the electrical box |

Note: When you replace a switch or outlet, you should not only shut off all power to the circuit, but also turn off all switches and unplug all lamps and appliances on it. Otherwise, the combined electrical demand of all the current-drawers turning on simultaneously might blow or trip the circuit interrupter.

## Switches

Among the many switch designs available at hardware and home-improvement centers are the common toggle switch, mercury (silent) switches, tap, push-button, and dimmer switches. All are replaced in one of two ways, depending on whether the unit has screw terminals or is backwired.

You can replace a defective wall switch with any other wall switch, but before you buy a replacement, consider the advantages of the different types available. A mercury switch does not make any sound when it is operated, which may be desirable in places such as a bedroom. Dimmer switches regulate the amount of light coming from the bulb, so they not only control the mood of your lighting, but can save on electricity as well. Push-button and tap switches can be operated by an elbow as well as a finger, so they might be desirable in a laundry room or anywhere you are likely to have your arms full of packages.

Having chosen the type of switch, install it in one of two ways.

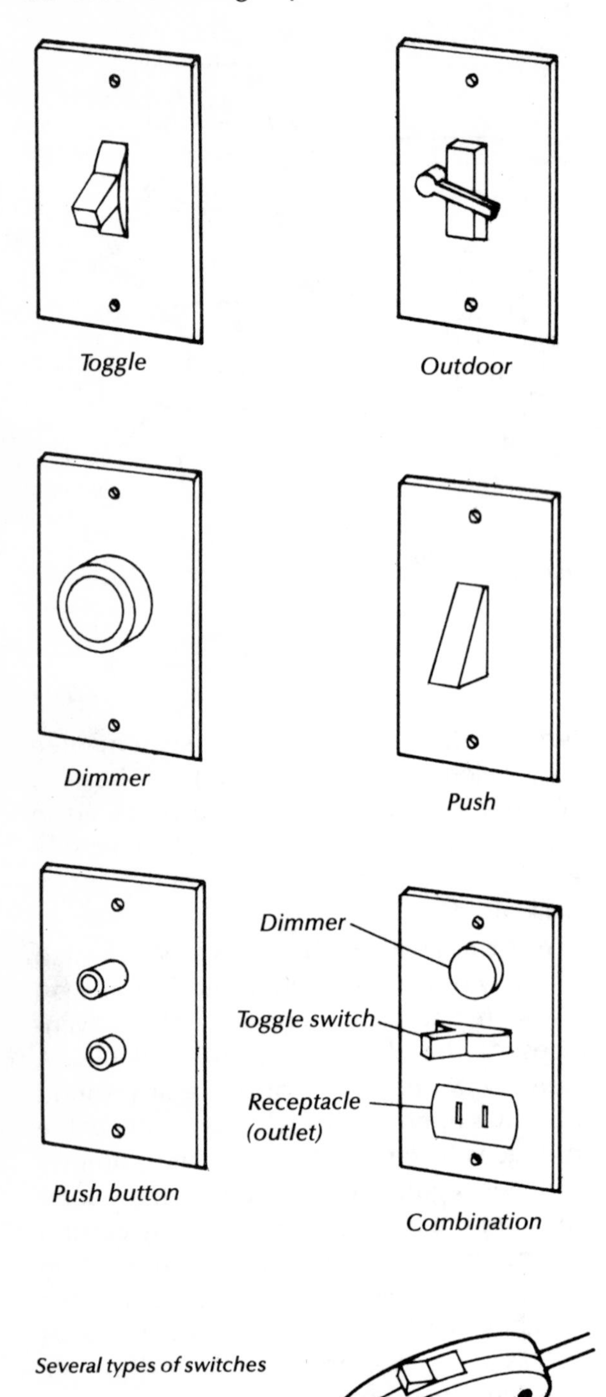


Several types of switches

## Replacing standard switches

Most of the switches sold today have color-coded screw terminals on their sides. Follow the coding as you install them.

1. **Remove the fuse or turn off the circuit breaker on every branch circuit that feeds the switch.**
2. Remove the two screws in the faceplate covering the switch and pry the faceplate away from the wall.
3. Remove the two mounting screws at the top and bottom of the switch.
4. Pull the switch out of its electrical box. The cable wires connected to the switch are heavy-gauge and stiff, but they will bend as you pull on the switch.
5. Depending on how the switch is connected, there could be as many as five wires attached to it, or as few as two. Even more confusing, you might discover that a white wire was painted black, or wrapped with black tape, and connected to a brass terminal (see page 35). Trust the electrician who installed the old switch. However the wires are connected, the old switch worked, so that is how you should connect your replacement unit. Make a diagram showing which wires are connected to which screw terminals. You can also tag each wire with a piece of tape, or, even better, if you have the space to work in, remove each wire one at a time from the old switch and place it on its corresponding screw terminal on the new switch.

   Loosen the screw terminals and detach the wires from the old switch. Hook each wire around its corresponding screw terminal in the new switch and tighten the loop with pliers, then tighten the screw.

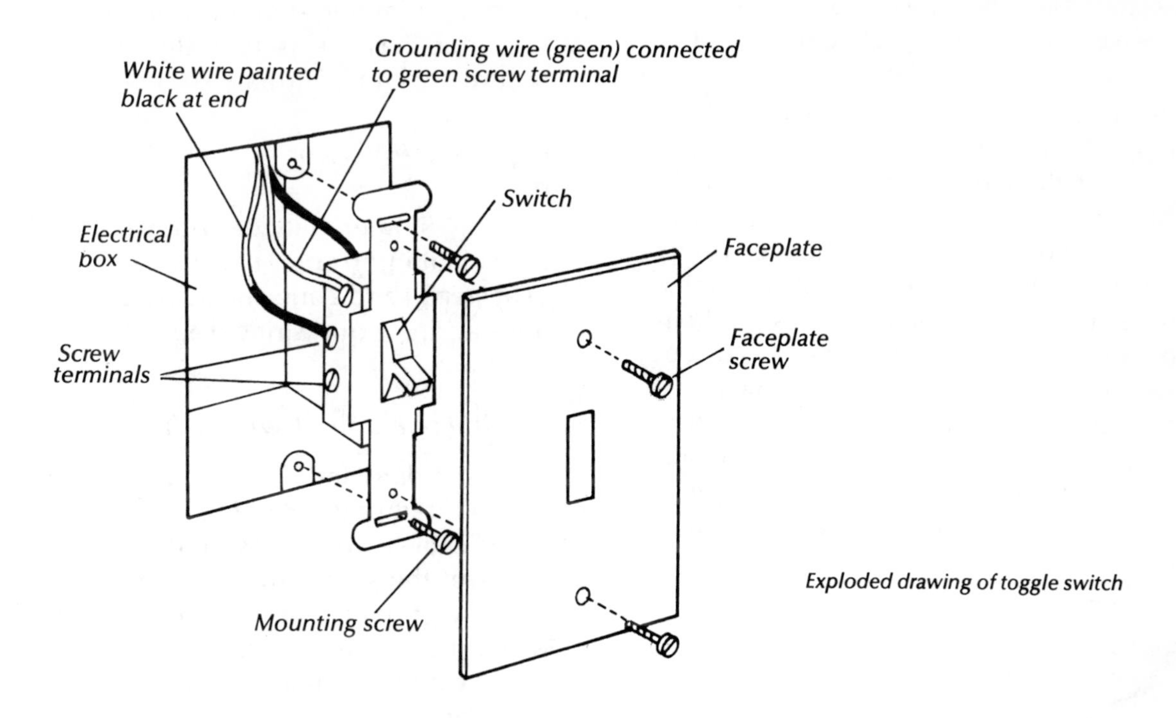


Exploded drawing of toggle switch

6. Push the switch into the electrical box.
7. Install the mounting screws at the top and bottom of the switch.
8. Install the faceplate over the switch and secure it with both screws.
9. Replace the branch circuit fuse or turn the circuit breaker on.
10. Test the switch two or three times. If it does not work, most likely one (or more) of the wires is on the wrong terminal. If the circuit interrupter blows or trips, one of the wires inside the electrical box is touching the box and shorting out the branch circuit. In either case, the remedy is to repeat steps 1–4 and check the connections against your wiring diagram.

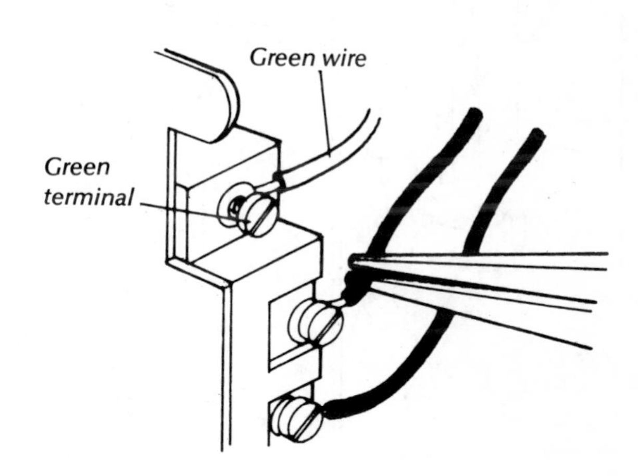


*Only the hot wires and the grounding wire are attached to a switch*

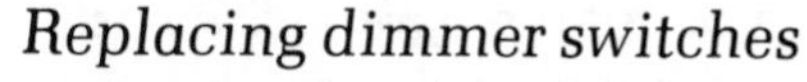

### *Replacing dimmer switches*

Dimmer switches come in a variety of designs and wattage ratings. Give yourself a safety margin by purchasing a dimmer that is rated higher than the total wattage of the bulbs it will be controlling.

1. **Remove the fuse or shut off the circuit breaker on every branch circuit to the switch you are replacing.**
2. Remove the faceplate and switch (see page 26, steps 2–5).

   Most dimmers have screw terminals similar to those of standard switches, and are connected to the house wiring system in the same manner. A few units have special wiring instructions on their packages, which must be followed exactly. Typically, the black cable wire is connected to the brass terminal and the white wire goes to the silver terminal. Hook the bared wires around their proper screw terminals and tighten them.

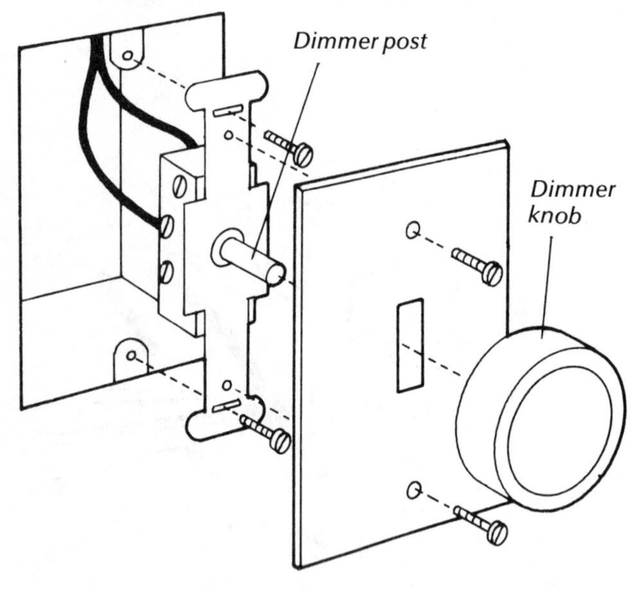


*Exploded view of dimmer switch. (The green wire has been omitted here and in following drawings for the sake of clarity.) One of the black wires is actually a white wire painted or taped black*

3. Push the dimmer body into the electrical box and secure it with its mounting screws.
4. Install the faceplate and tighten its mounting screws.
5. Push the dimmer knob onto its post. If the knob has a setscrew, tighten it to secure the knob. Turn the knob to its Off position.
6. Replace the fuse or turn on the circuit breaker.
7. Test the dimmer. If it does not work or if the circuit interrupter blows or trips, repeat steps 1 and 2 and check the connections against your wiring diagram.

### *Replacing backwired switches*

Some standard and dimmer switches are backwired, rather than having screw terminals. The cable wires are inserted in holes in the back of the switch, where they are automatically gripped and held in place. The holes are labeled, and often color-coded as well. A backwired switch can replace almost any standard switch.

1. **Remove the fuse or turn off the circuit breaker on every branch circuit feeding the switch.**
2. Take off the faceplate and remove the old switch. If it is a backwired unit, you will find a tiny wire-release hole or slot near each of the wire holes. Push the blade of a screwdriver or a piece of wire (such as a straightened paper clip) into the wire-release hole and simultaneously pull on the wire. It will come out of its hole.
3. Straighten the ends of the cable wires and clean them until they are shiny; scrape them with the back of a knife blade, or rub them with steel wool or sandpaper.
4. A stripping guide is printed on the back of the backwired switch's casing. Strip off insulation or cut the bared wire ends to the exact length indicated by the stripping guide.

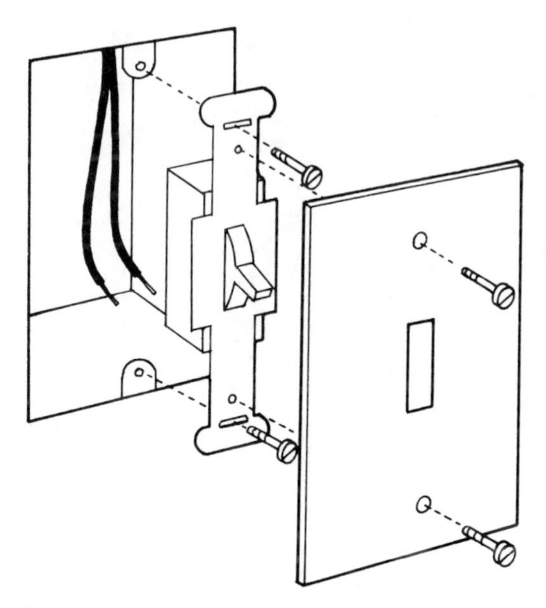

The front of a backwired switch—note the absence of screw terminals

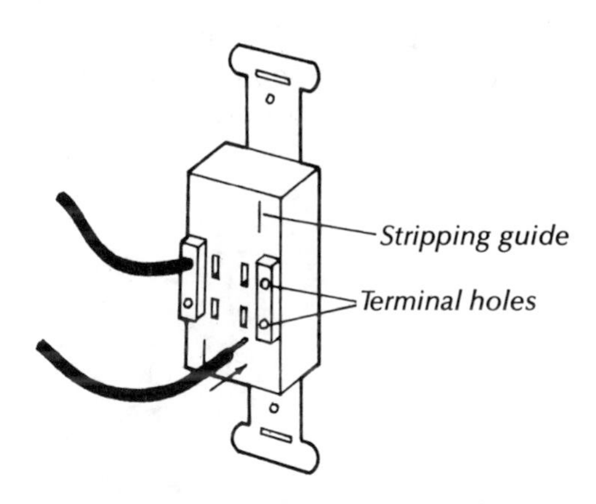


The wires are inserted in terminal holes

5. Following the coding or color guide printed on the back of the switch, insert the wires in their proper holes and push them firmly into place.
6. Push the switch into the electrical box and secure it with its mounting screws. Install the faceplate.
7. Replace the fuse or turn on the circuit breaker.
8. Test the switch. If it does not work, repeat step 1, take off the faceplate and the switch's mounting screws, and examine the back of the switch to be certain each wire is in its proper hole according to the color code. If you have connected the unit properly and it still does not work, the switch is defective and should be returned to the store.

## Outlets

If the fuse blows or the circuit breaker trips every time you plug in an appliance, or if you plug in an appliance and it doesn't work, most likely the outlet is faulty and must be replaced. While there are several types of outlet designs, most of them are installed in the same manner, although the location and number of wire connections may differ considerably.

A new ruling by the National Electrical Code states that an old outlet may be replaced only by a grounding-type unit, which can accept both two- and three-prong plugs (see page 55). Grounding receptacles have two slots to receive the prongs of an appliance plug; one of them is slightly shorter than the other and is connected to the hot side of the cable—that is, the black wire. Grounding is provided by a

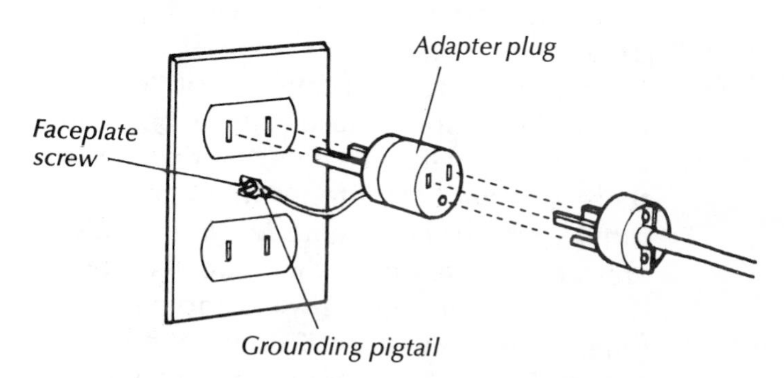


An adapter can convert an old-style receptacle. The pigtail must be connected to the faceplate to insure grounding protection

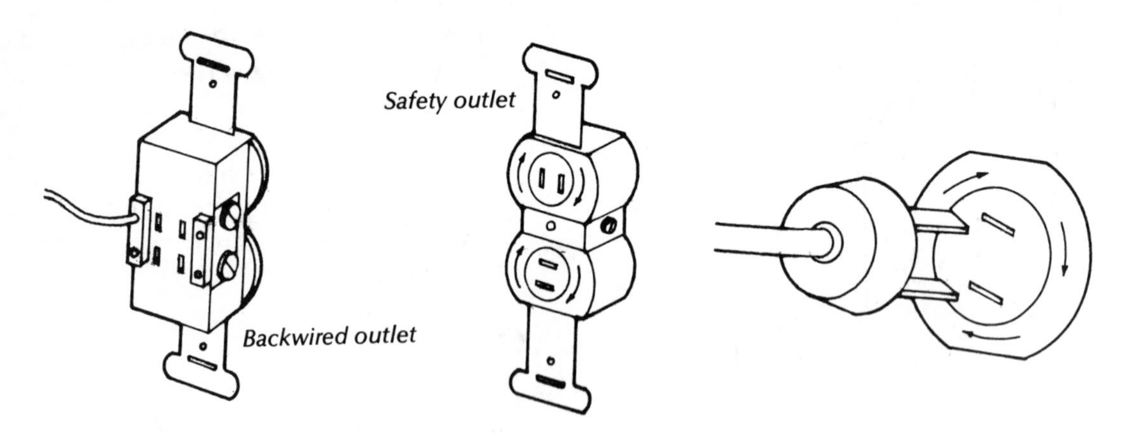


Backwired and safety outlets

hole beneath and between the slots, designed to accept the third, grounding prong of three-prong plugs; it leads to a green screw terminal on the receptacle. A wire is run from the green screw to the electrical box, to further protect against shock.

There is no reason you cannot change the style of outlet you are replacing, and there is quite a variety to choose from. If you are replacing a single-receptacle outlet, consider using a double-receptacle unit. In a child's room, you might want to consider a safety outlet, which has a plastic disk that must be rotated before a plug can be put in the receptacle—a task that demands more skill and strength than small children usually possess.

When you are replacing any outlet you are liable to open the electrical box and discover a jungle of wires, particularly if there are two receptacles next to each other, or if the two receptacles in the outlet are fed by different branch circuits. Then there are the grounding wires and perhaps jumper wires (see page 33) . . . so the best approach in making your replacement is to carefully diagram where each wire comes from and which screw terminal it is attached to. Better still, tag each line—and even with that, transfer the wires one at a time from the old receptacles to the new. If you should make a mistake in your transfer, nothing will happen—literally. You will not get any electricity into the receptacles, and any appliance you plug into them will not work.

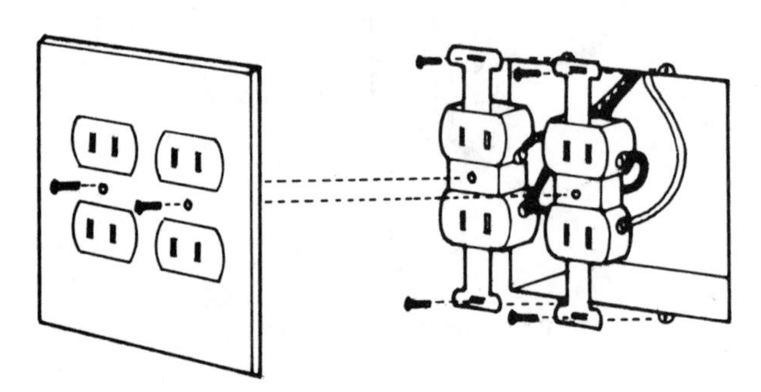

A multiple outlet can contain a multitude of wires. Diagram them carefully before removing the outlet

## Replacing an outlet

1. **Remove the fuse or turn off the circuit breaker on every branch circuit servicing the outlet. Make certain that ALL of the circuits entering the outlet box are shut off.**
2. Remove the faceplate.

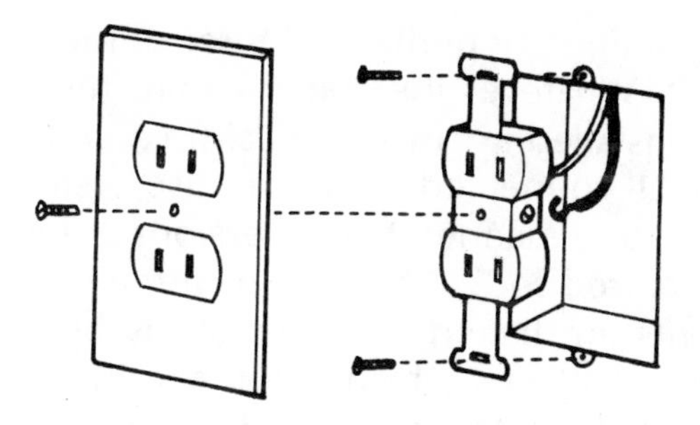

*Exploded view of a standard outlet*

3. Remove the outlet's mounting screws and pull the outlet out of the electrical box.
4. Look at the wires very carefully and diagram their connections.
5. Remove one wire from the old receptacle and place it around the corresponding screw terminal of the replacement unit. Tighten the terminal screw. Transfer the wires, one at a time, until the new unit is completely connected and you can remove the old one.
6. Connect a 2″–3″ length of bare copper wire (purchased separately) to the green screw terminal. Attach the free end of the wire to the electrical box, using a screw (purchased separately) inserted in any of the screw holes in the sides of the box, or a grounding clip that fits over the edge of the box.

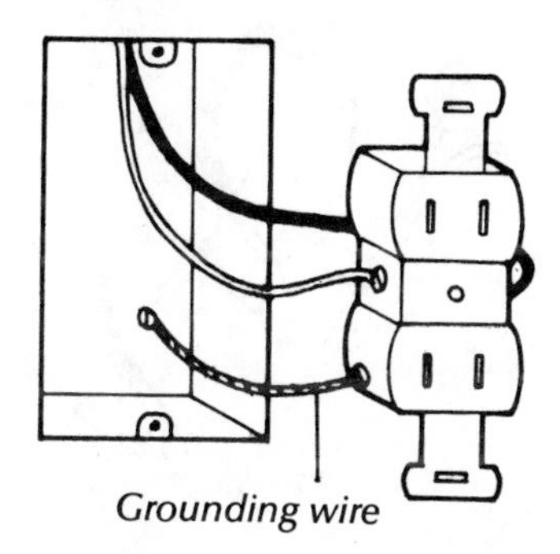


*The grounding wire is connected to a screw terminal and to the electrical box*

7. Push the outlet into the electrical box and tighten its mounting screws.
8. Install the faceplate.
9. Replace the fuse or turn on the circuit breaker.
10. Test the receptacles by plugging in an appliance or lamp. If the circuit interrupter blows or trips, a wire may be touching the inside of the electrical box. Repeat steps 1–3 and check the connections against your diagram.

### *Replacing backwired outlets*

1. **Remove the fuse or turn off the circuit breaker on every branch circuit serving the outlet.**
2. Remove the faceplate.

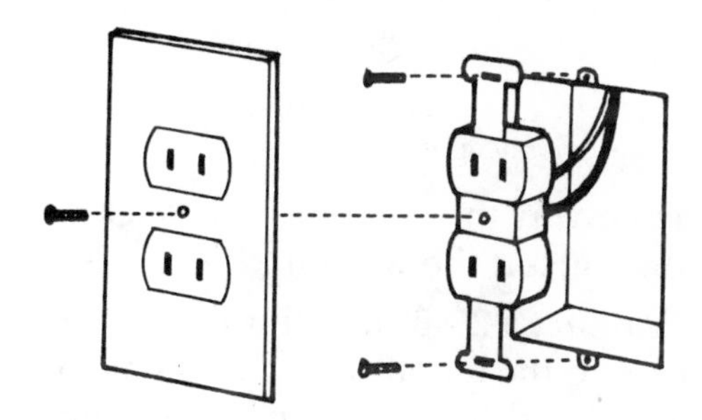

*Pull the outlet out of its box*

3. Remove the mounting screws holding the outlet and pull the unit out of the electrical box.
4. Diagram the wiring connections.
5. Push the blade of a screwdriver or a piece of wire into the wire-release holes and pull each wire out of the unit.
6. Straighten the wire ends.
7. Measure the wire ends against the stripping guide printed on the new unit. Strip more insulation or trim the wires to the proper length.

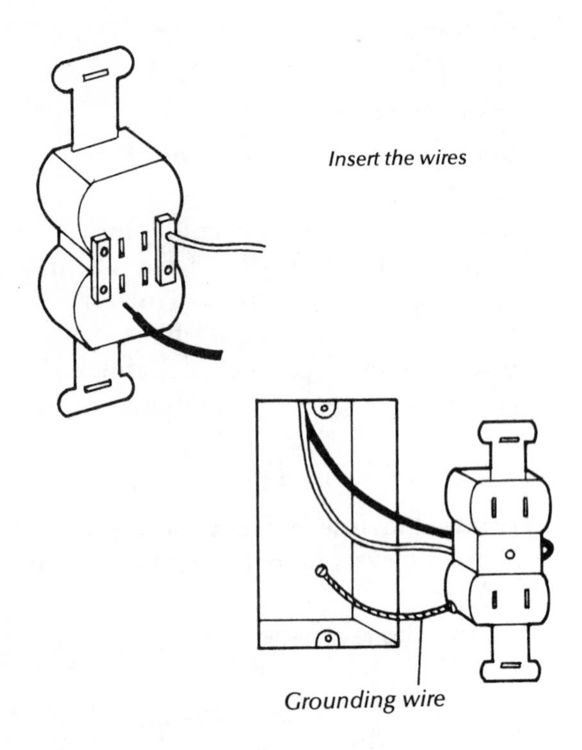


*Insert the wires*

*Connect the grounding wire*

8. Following the coding on the back of the unit, insert the wires in their slots.
9. Connect a 2″–3″ length of bare copper wire to the green screw terminal and to the electrical box (see page 31, step 6).
10. Push the outlet into the box and secure it with its mounting screws.
11. Install the faceplate.
12. Replace the fuse or turn on the circuit breaker.
13. Test each receptacle by plugging in an appliance or light. If the appliance or light does not operate, shut off all electricity to the box, remove the outlet, and check the wiring against your diagram.

## Combination Switches

Combination switches can consist of a switch and a receptacle, a switch and a night-light, or perhaps all three in the same unit. Whatever the combination, they can be installed in any electrical box, and they are all connected to your house system in the same manner. Some versions are manufactured so that the two or three components are linked by metal strips bridging their contacts, which automatically allows all of them to operate off a single cable entering the box. Or, if there is more than one circuit in the box and you want the components to be fed by different branch circuits, you can connect a separate cable to each component and break off the metal links so that there is no crossover of elec-

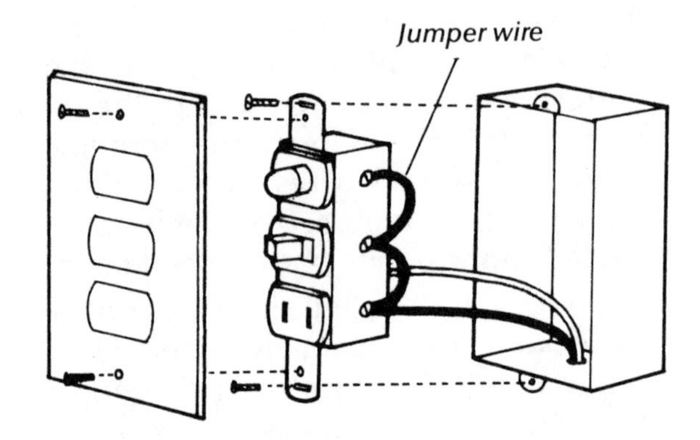


*Jumper wires can connect separate components in a combination unit*

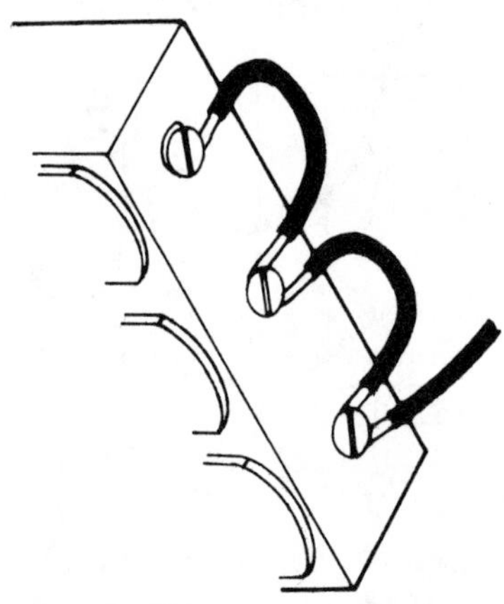

*Strip the jumper wire where it will be wrapped around a screw terminal*

tricity within the unit. Other combination units have completely separated components, so if you want the separate parts to operate on the same circuit it is necessary to attach jumper wires between their terminals.

*Jumper wires*—A jumper wire must carry the same amount of electricity as the cable feeding into the box, so it should be at least the same wire gauge as the cable, which most likely means #12 or #10 wire. To make a jumper wire, simply open a piece of cable and cut off 2″–3″ of both the black and white wires. Strip ½″ of insulation from each end of the wires and bend the bared ends into loops that can fit around the unit's screw terminals. If the jumper must carry electricity from the receptacle to the switch and night-light in the same unit, you can remove a ½″ length of insulation from around the center of the wire and then bend the wire so that the stripped section loops around the intermediate terminal.

### *Installing combination switches*

1. **Remove the fuse or turn off the circuit breaker on every branch circuit entering the electrical box.**
2. Remove the existing switch (see pages 26 and 28) or outlet (see pages 30 and 31).
3. If jumper wires are to link the components of the unit, you will need a black wire for the hot side of the unit and a white wire for the return side. Strip approximately ½″ of insulation from the ends and middle of each wire. Loop the jumper wires around the screw terminals on the sides of the unit. One end of each wire must be attached to a terminal that will also accept the branch circuit cables in the electrical box.

   If no jumper wire is needed, connect the black cable wire to one of the brass terminals and the white wire to one of the silver terminals.

   If the components are to be fed by different cables, make certain that you put the black and white wires from each cable on the same component.

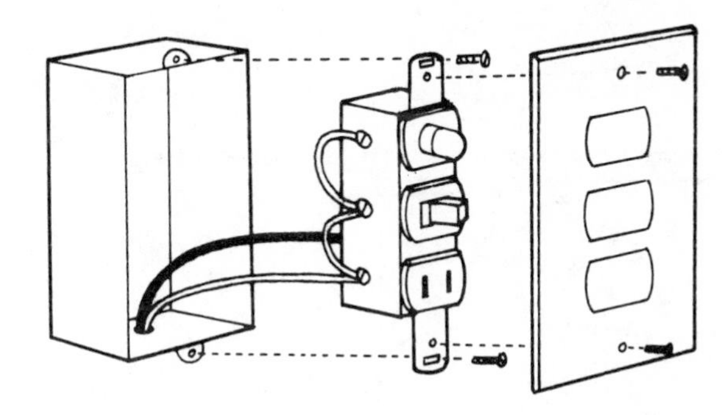

*Installing a combination unit. The white and black wires are connected the same way*

4. If the unit has a green screw terminal, connect a bare grounding wire to the terminal and attach it to the electrical box or to the back of a mounting screw (see page 31, step 6).
5. Push the unit into its box and secure it with mounting screws.
6. Install the faceplate.
7. Replace or turn on the circuit interrupter and test each component in the unit. If any part of the unit does not work, you have most likely attached a wire to the wrong terminal. Shut off the electricity to the box, remove the unit, and check all of your connections.

## REPLACING CEILING LIGHTING FIXTURES

Replacing a ceiling fixture is often the only home electrical repair anyone ever has to make. The difficulty in installing any new fixture is more likely to be in securing it to

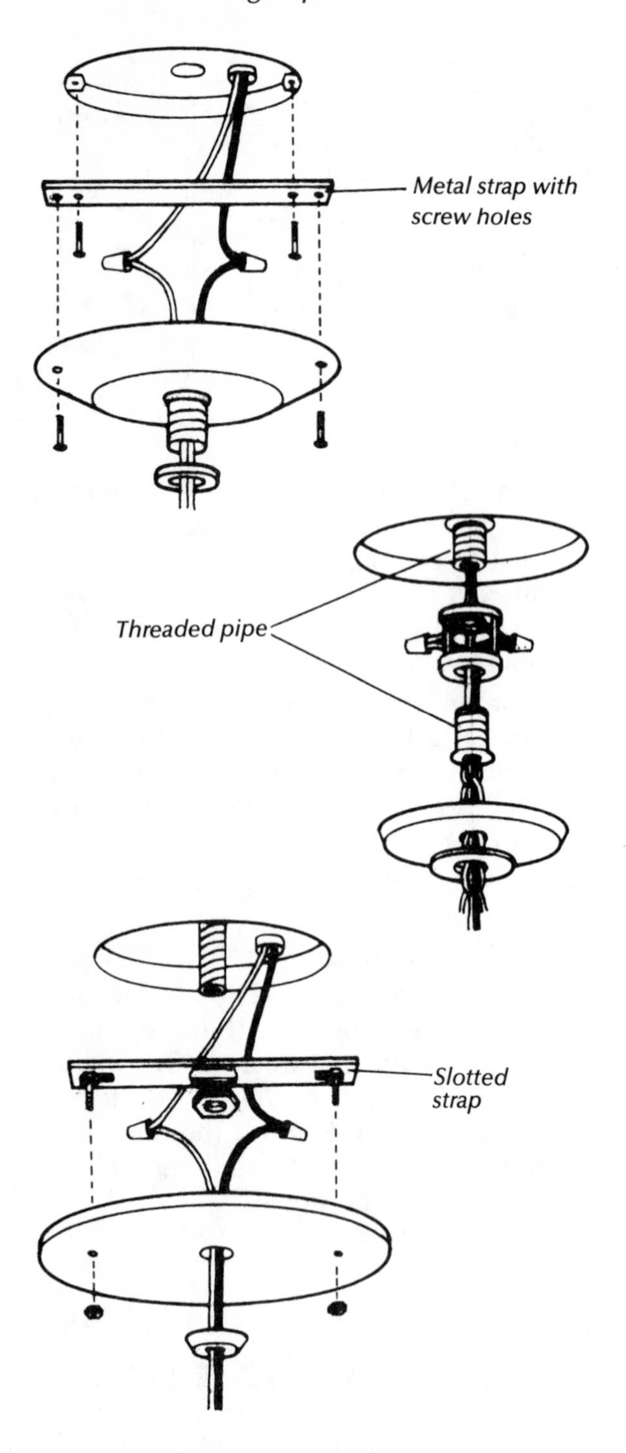


Three ways of hanging a fixture from its ceiling box

the ceiling, rather than how its wires are connected to the house electrical system. There are a variety of ways in which ceiling fixtures are held in place under the round or octagonal box that contains the wires that feed it. The box may have a threaded pipe hanging down from its center, which fits through a hole in the base of the fixture, which is held to it by a nut. Unfortunately, the pipe may be any of several diameters, and it may not fit into the new fixture. Remove it from the electrical box simply by unscrewing it.

There are also several different-size metal brackets that can be attached to the pipe or to the box by setscrews or bolts. They have narrow slots down their middles to receive bolts that can be slid back and forth until they align with holes drilled through the base of the fixture.

Whatever attaching mechanism you find in the ceiling box, it can be changed to suit the mechanism that accompanies the fixture, but you may have to make a trip or two back to the lamp store before you get the correct pieces of metal. Determine exactly how you will hang the fixture before you proceed to make its wire connections.

1. **Remove the fuse or shut off the circuit breaker on every branch circuit entering the fixture box.**
2. Any parts of the fixture that you can remove should be taken off before you loosen its base from the ceiling. This includes shades, bulbs, perhaps a chain—whatever.
3. Loosen the base of the fixture from the ceiling box. If it is held in place by a hollow center pipe that holds the base with a nut, the wires from the fixture lamp socket(s) probably come from the box down through the pipe. Or there

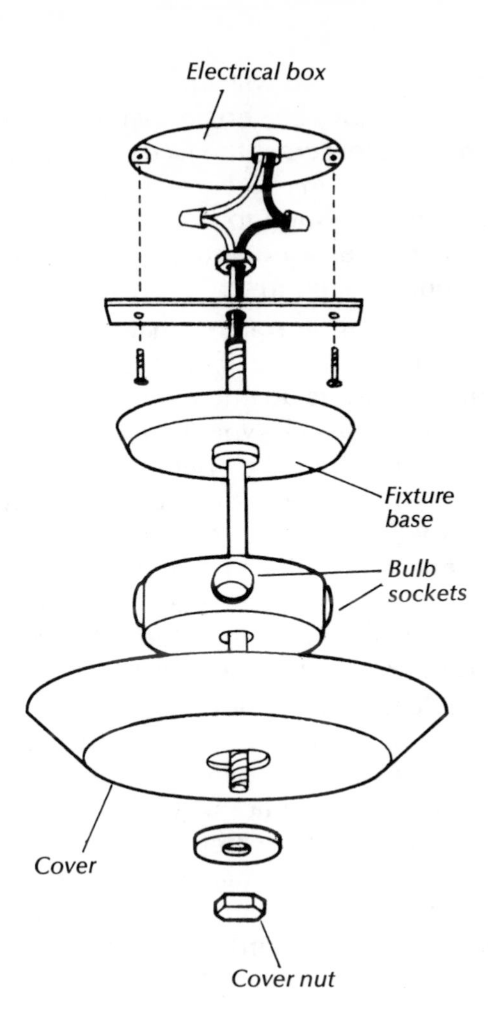


*Disassemble the fixture*

may be a cover over the baseplate that is held in position by screws or bolts that extend down from a metal bracket attached to the ceiling box. Whatever nuts, bolts, or screws have to be loosened, when you are finished, the fixture should end up hanging by its wires, so don't let go of it—the wire connections may not be strong enough to hold its weight.

4. With the base of the fixture held away from the electrical box, pull the cable

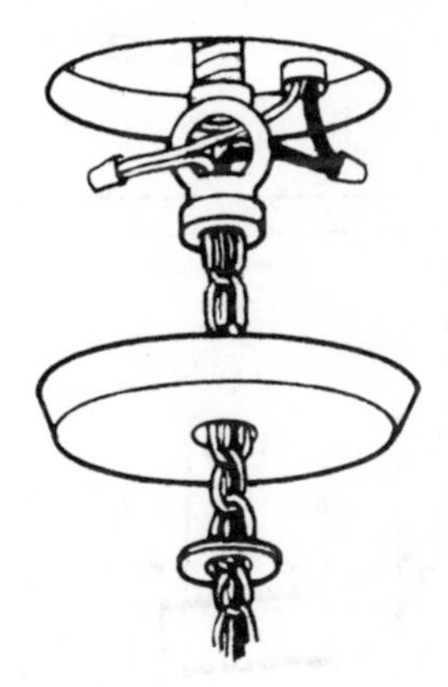

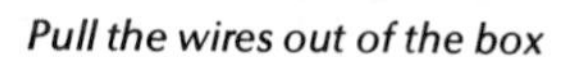

*Pull the wires out of the box*

wires down out of the box. Carefully note which wires from the fixture are attached to which cable wires. In general, you should find that all of the black wires are together, and all the white wires are also connected with a single wire nut or tape. However, if a wall switch controls the fixture, you will find a black wire from the switch connected to the black wire of the fixture; the white wire from the switch is connected to the black wire in the circuit cable, and has been painted black or wrapped with black tape at the end; and the white wire from the fixture is attached to the white wire in the circuit cable. Be sure you fully understand where the wires are connected before you detach any of them. If possible, get an assistant to diagram the wiring as you inspect it.

5. Remove whatever wire nuts contain the wires from the fixture and pull the fixture wires free. At this point you can put the fixture down somewhere out of your way.

6. Assemble the replacement fixture, or do whatever has to be done to prepare it for hanging. You have to figure out where its wires will enter the electrical box—they

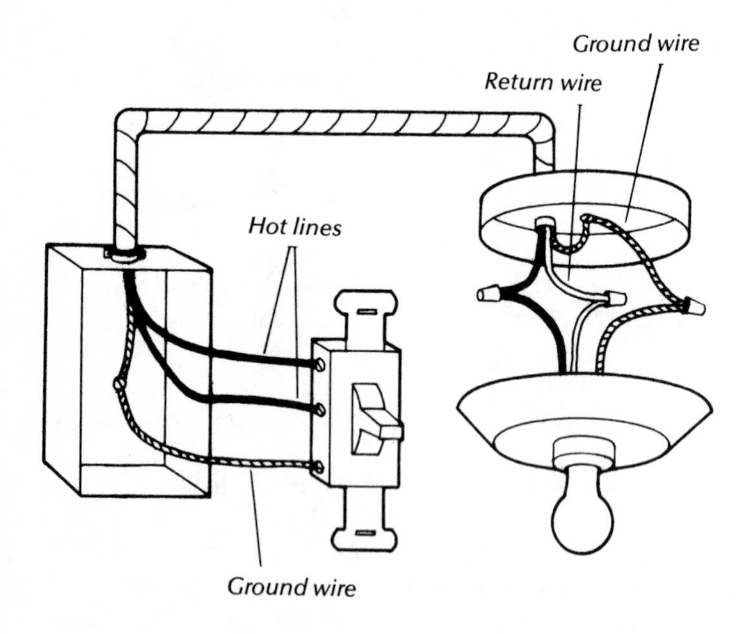


*How a switch is connected to a ceiling fixture*

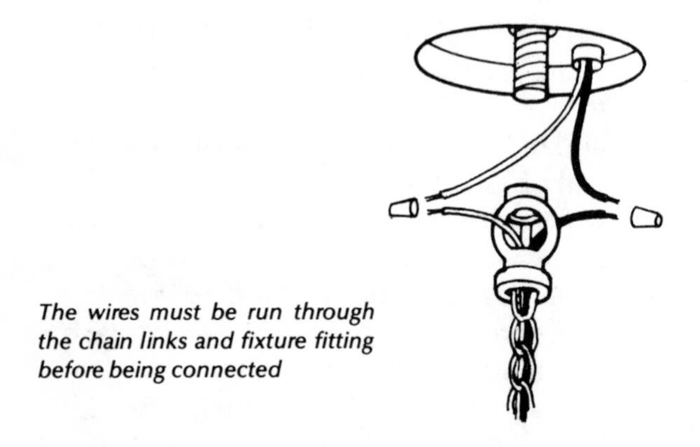

*The wires must be run through the chain links and fixture fitting before being connected*

may have to be threaded through the links of a swag lamp's hanging chain, or up through the pipe in the electrical box, or any number of other possibilities. Unless the fixture is completely ready for installation or is extremely simple, the manufacturer has probably provided instructions to guide you. Also check out the system that will support the fixture. For example, be certain that the bolts of the bracket across the face of the electrical box are in the correct position and are long enough to hold the base, and that the pipe is not too large or too small to hold the fixture.

7. When you are sure that all of the mechanics of the fixture are in order, bring its wires up into the electrical box. You have to make your wire connections in the electrical box before you can secure the fixture against the ceiling, and you will need both hands to make the connections. If you are standing on a high stepladder, you may be able to rest the fixture on its top step. Otherwise, you may need an assistant to hold the unit for you while you do your electrical work, or perhaps you can tie the fixture to the electrical box with wire or cord. Also bear in mind exactly what the wires are supposed to do before they enter the box. If the unit hangs from a chain, for example, the wires should weave through every third or fourth link and enter the electrical box through the center pipe. You have to do the weaving *before* you make any connections.
8. Connect the black and white wires from the new fixture to the circuit cable wires. If the unit has no white wire, there will be a white or light-colored tracer thread enclosed in the insulation of one of the wires; that wire is connected to the white wires in the box. The plain black wire goes with the black cable wires. Twist the fixture wires around the appropriate wires in the box and screw wire nuts down over the connections, or wrap the bare wire ends with electrician's tape. Then push the wires up into the box.

9. Position the base of the fixture up over the electrical box and secure it to the bracket or pipe attached to the box.
10. Replace or turn on the circuit interrupter and test the light. If it does not work, you have made an incorrect connection and will have to recheck the wiring.

*Concealing fixture boxes*

Ceiling or wall fixtures can be removed from service altogether, but that leaves you with an open electrical box. After the fixture has been taken off, cover the bared wires in every connection with wire nuts, or wrap the wire ends with tape. Push the wires back into the box and screw a standard metal cover plate to the face of the box. Paint or wallpaper over the cover plate; a thin coating of plaster or wallboard compound applied first will make it more inconspicuous.

# REPAIRS TO INCANDESCENT LAMPS

Incandescent lamps are used in ceiling and wall fixtures as well as freestanding lamps. No matter how they are used or what they look like, they all consist of a socket and a switch, which are wired either directly to the house circuit or to a cord and plug inserted in an outlet.

When a lamp fails to operate, most often the problem is with the switch or the socket. If these are in working order, the only areas left to examine are the cord and its plug. The sockets used in all lamps are virtually identical, with only a few variations in their designs. Most notably, some switches operate via a pull chain while others use a rotating knob or push button. Switches and sockets are difficult and time-consuming to repair, so nobody ever fixes them; they are simply replaced with a unit purchased at any hardware store, five-and-dime, housewares or home-improvement center, at the cost of something around a dollar.

Sockets are all manufactured to the same specifications, so they will all fit into whatever fixture or lamp you are repairing. You can replace a pull-chain-switch socket with one with a knob, and some sockets are constructed so that the switch makes three different contacts, allowing the use of a three-way bulb. Depending on what the socket is fitted into, all or some of its outer shells may be used. Ceramic-based ceiling fixtures, for example, form a shell by themselves, so only the actual socket and its switch are placed inside the ceramic.

## Taking Lamps Apart

Don't let the looks of any lamp fool you. They are all assembled in exactly the same way, no matter what materials have been used to make them or what they look like.

1. **Unplug the lamp.**
2. Turn the unit upside down and peel off the pad glued to its base.

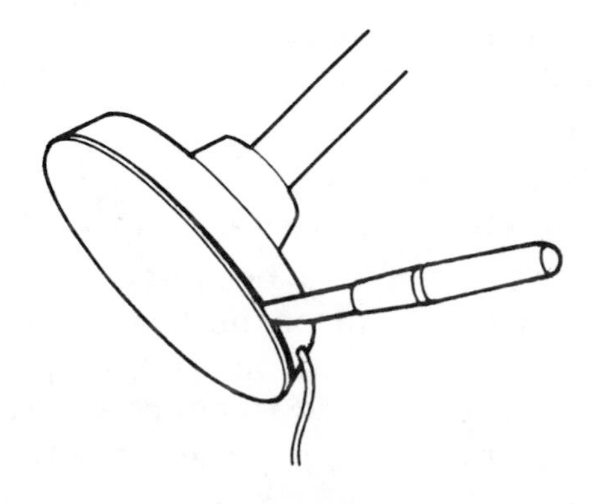

*Peel or pry off the base pad*

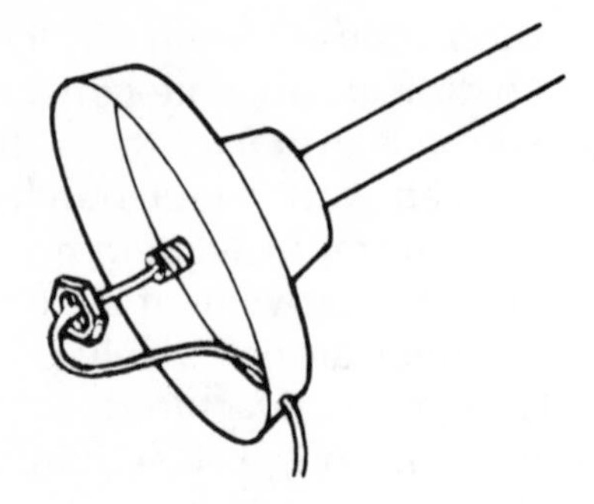

*Undo the nut at the bottom of the center pipe*

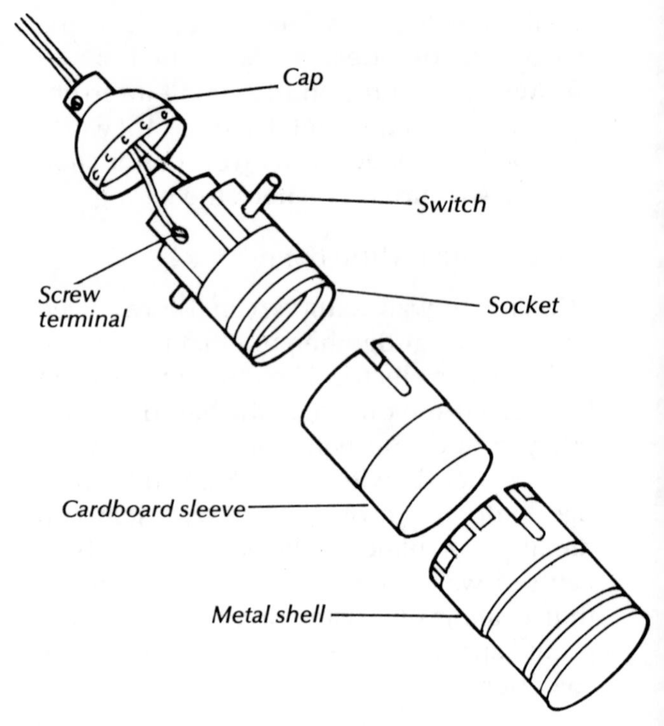


*Anatomy of a light socket*

3. Undo the nut around the bottom of the pipe or rod in the center of the base. You can remove all of the exterior parts of the lamp by pulling them off the pipe.
4. The socket may be screwed to the top of the pipe or it may have a special bracket. The socket has a metal exterior shell that fits into a cap. Twist the shell slightly and pull it out of the cap. Slide off the cardboard sleeve that protects the socket.
5. The wires in the cord are connected to screw terminals on either side of the socket switch. Undo the screw terminals and remove the cord. Pull the cord out of the center pipe if you plan to change or repair it.
6. Unscrew the cap from the pipe. If the socket/switch is defective, take it to the nearest hardware store and buy a similar unit to replace it.
7. Pull the new socket off its cap and remove the outer shell and paper tube.
8. If the old cord is to be reused, it should be sticking out of the lamp pipe. To install a new cord, push it through the pipe, leaving a few inches of wire sticking out of the top of the pipe. Thread the cord through the center of the socket cap and attach the cap to the end of the pipe.

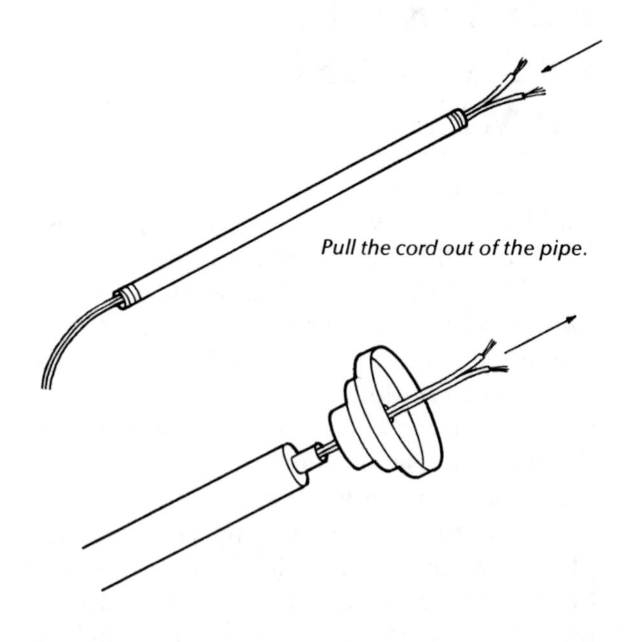

*Pull the cord out of the pipe.*

*Pull the new cord through the pipe and the socket cap*

9. Strip about ½″ of insulation from the ends of the cord wires. With a knife, cut through 2″ or 3″ of the insulation between the two wires and pull them apart. Each wire should remain completely insulated.
10. Twist the strands of each wire clockwise (solder them if you like) and wrap one wire around each of the terminal screws.
11. Insert the socket in the cap. Slide the cardboard sleeve over it, and twist the outer shell into the cap.
12. Thread the components of the lamp onto the pipe and tighten the base nut.
13. Glue the base pad to the bottom of the lamp with white or cream-colored glue.

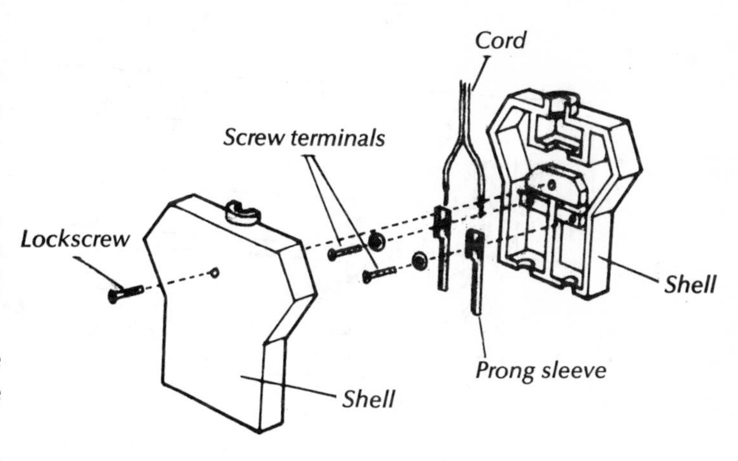


*Some plugs come apart so that cord wires can be inserted in crimp-on terminals*

## Replacing Plugs

Damaged plugs can be replaced by any of several UL-listed plugs. Some plugs are designed to hold unstripped wires in a slot with metal prongs that poke through the insulation to touch the wires. Other plugs have terminal screws at the base of their prongs. To secure a cord to a screw-type plug, thread the cord through the hole in the center of the plug, then pull the wires apart and tie them in an electrician's knot to provide strain relief when the cord is yanked out of a receptacle. Once the knot is tightly nestled between the prongs, strip ½″ of insulation from the wire ends and hook them clockwise around the plug terminals. Slide the plastic or cardboard disk down the prongs to cover the terminals.

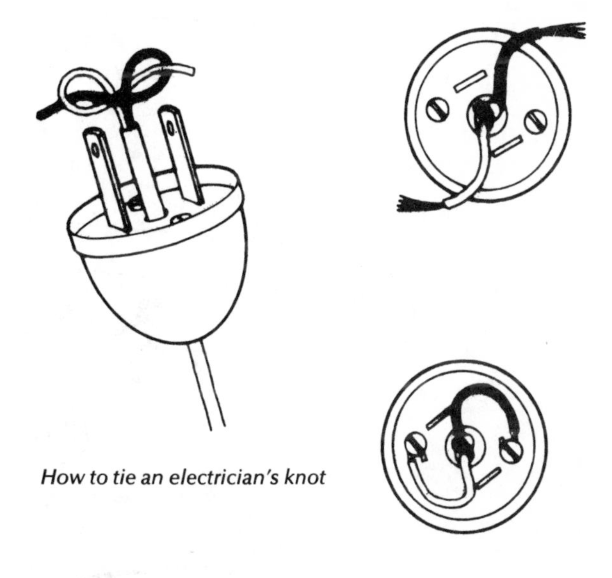

*How to tie an electrician's knot*

# FLUORESCENT FIXTURES

Fluorescent fixtures consist of a tube or tubes, lamp holders (sockets), a ballast, and perhaps a starter.

*Tubes*—Fluorescent tubes can be as large as 1½″ in diameter and 8′ in length; they are also made in U and circular shapes. They come in a variety of "white" colors; "cool white" (sometimes called "daylight") is bluer than "warm white." Fluorescent tubes last a long time compared to incandescent bulbs. Because they consume more electricity when they are starting up

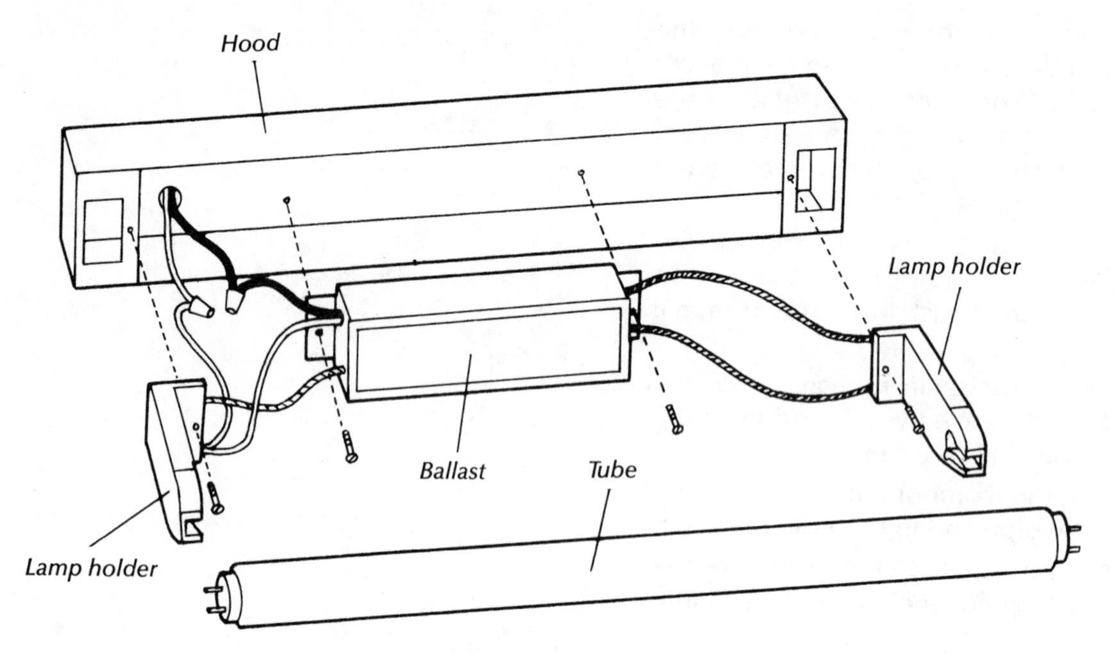


Exploded view of rapid-start fluorescent light

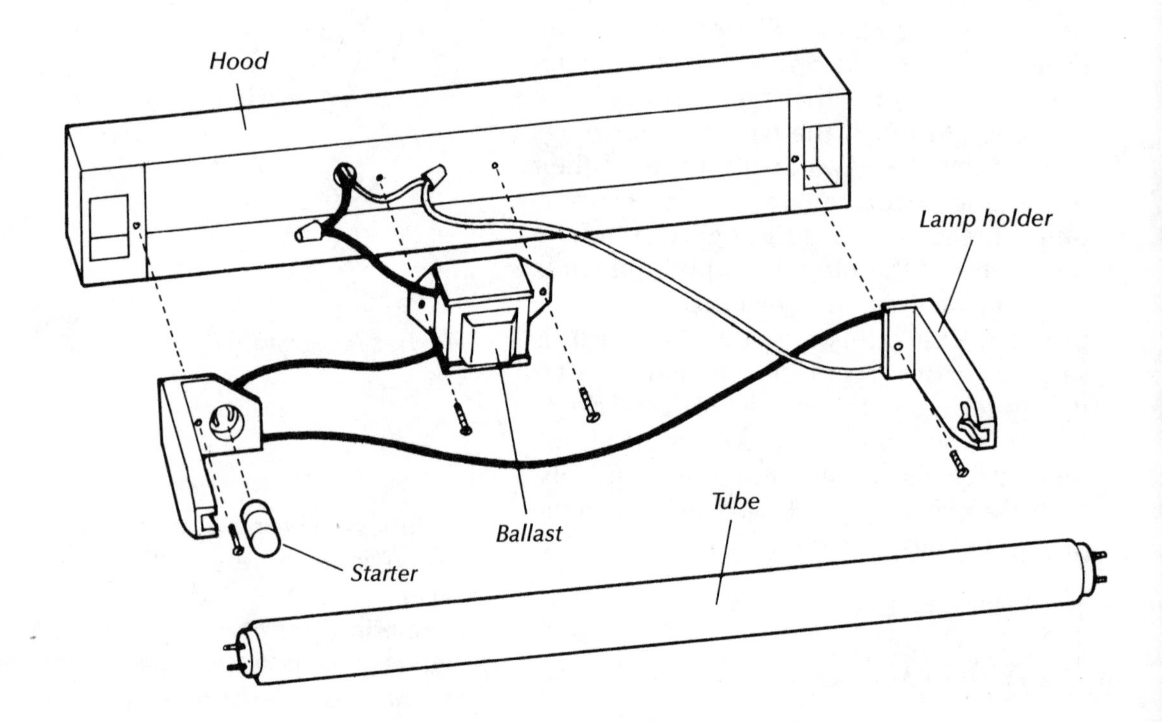


Starter-type fluorescent light

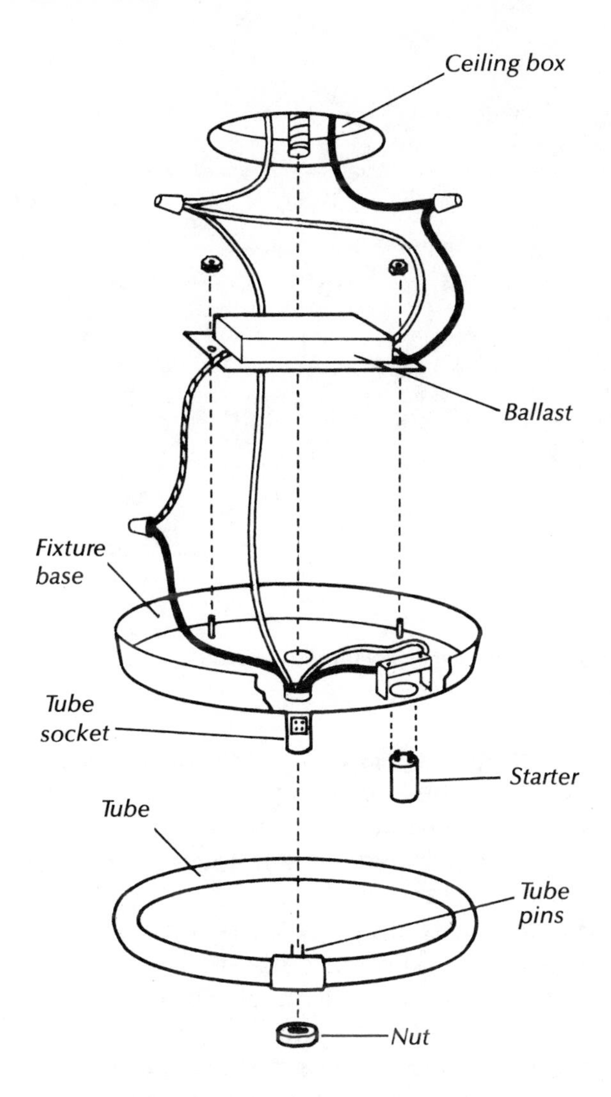


Circular starter-type fluorescent fixture

than afterward, if you plan to turn a fluorescent light on again within 15 minutes, generally it is better to just leave it on.

Most problems that occur with fluorescent fixtures involve the tube. If the tube fails to start, or starts slowly, it is defective; if the light is dim or blinking, the tube has nearly burned out. You can continue using it, but have a replacement available. In a two-tube fixture, replace both tubes at the same time. Because they are wired in series (see page 113), an old or defective tube will impair the efficiency of a good tube. Buy the type and wattage specified on the unit's ballast sticker. The type will be given as "start," "rapid-start," or "instant start."

*Lamp holders*—These are ceramic tabs that stick out of the hood of the fixture. They have slots to receive the pins in the ends of straight tubes. Rotate the tube 90° until its pins lock in the slots.

If the lamp holders are not correctly

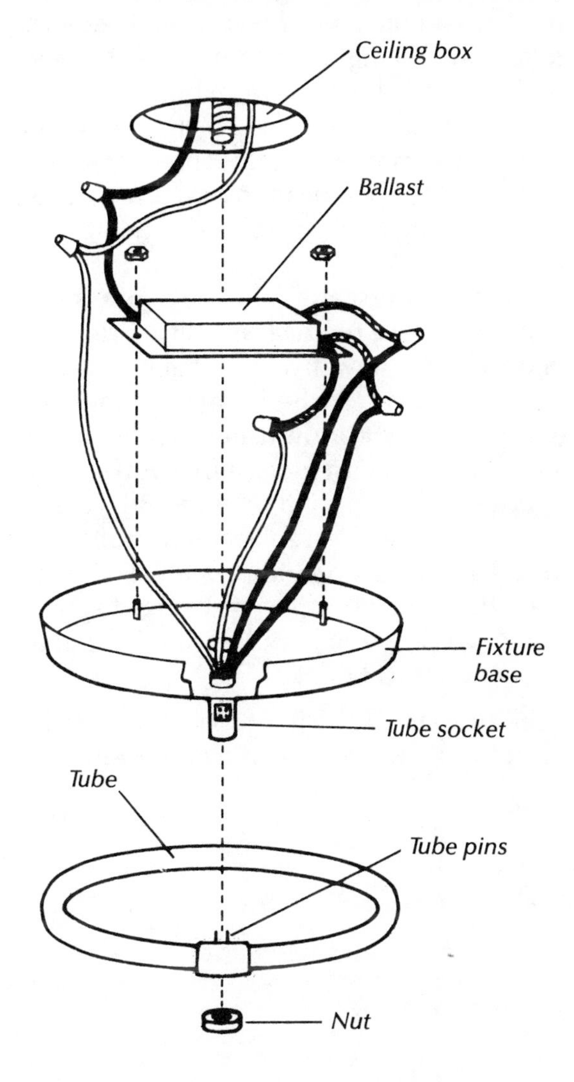


Circular rapid-start fluorescent fixture

situated in the fixture, the pins in the tubes will not make full contact, and the light will blink on and off. If a holder breaks, you can remove it by loosening the screw at its base, then disconnecting the pair of wires connected to it. Take the old holder with you to the store to be sure you get the proper replacement.

*Ballast*—This is a small, heavy metal box that is actually a kind of transformer (see page 112).When the light is first turned on, the ballast increases the voltage it receives to activate the light. Then it limits the electricity to keep the tube from burning out. A sticker on the side of the ballast provides a wiring diagram for connecting it to the fixture, plus specifications for the fixture, including the type of replacement parts to buy.

If your fluorescent light has a loud hum, you can replace the ballast with an identical unit or get a new fixture (which will cost about as much). If the humming does not bother you, continue using the fixture so long as the ballast does not overheat, as indicated by the light dying. Be careful when you install a replacement ballast. It must have the proper rating and be wired to the fixture according to its wiring diagram. If it is not properly installed, the lamp may not work, or will blink.

*Starter*—In older fluorescent lamps the starter is a little metal can about an inch or two long that has two screw terminals attached to one end. To install a starter, push it into its socket in the fixture hood and twist it clockwise until it locks. Modern instant-start and rapid-start fluorescents get the electricity they need directly from the ballast and do not have starters.

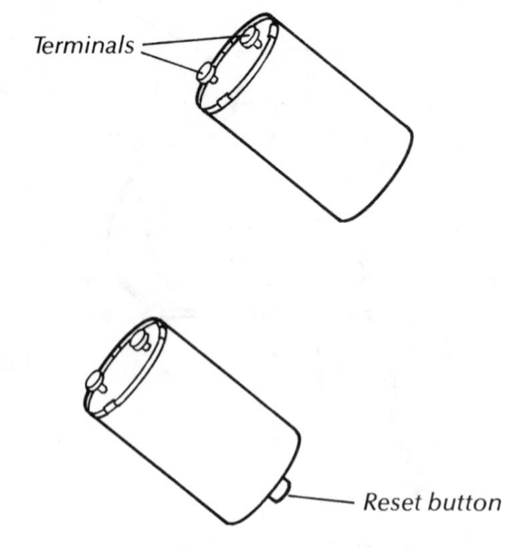


*Starters. Some have a reset button*

Starters occasionally burn out and need to be replaced. If the ends of a tube light up but the middle remains dark, or if the ends remain lit after the fixture is turned off, or if the light comes on only occasionally, the starter is defective. Be sure to replace it with a starter of exactly the same rating.

# CHAPTER 3 • Preparing to Upgrade Your House Electrical System

The basic repairs that must occasionally be made to a house electrical system do not require a knowledge of electrical theory, or any particularly technical information other than that discussed in Chapters 1 and 2. But the ultimate repair to a house system is upgrading its electrical capabilities, and before you tackle *that* you ought to have at least a passing acquaintance with the world of electrical engineering.

The electrical experts, like specialists in every field, have devised their own peculiar language, both verbal and written. Some of their terms are a part of our everyday language, such as:

*Ampere*—Amperage is speed, or the rate of electrical flow. One ampere is the number of electrons that pass a given point within one second when they are under pressure of one volt and the conductor (the wire) has a resistance of one ohm. Wires, circuit interrupters, and electrical equipment are always rated according to how many amps they can carry without being overloaded.

*Volt*—Voltage is pressure. Electricity moves through a wire only when it is pushed; the voltage tells you how hard a push it is getting. One volt is the pressure needed to move one ampere through one ohm of resistance.

*Ohm*—Every wire creates some resistance to the electrons moving through it. That resistance is called "ohmage," and how much of it exists in a given wire depends on the wire's length and diameter, and the metal of which it is made.

*Watt*—Wattage is the amount of work done; in other words, watts = total power.

Two formulas express the relationship

between amps (I), volts (E), ohms (R), and watts (W):

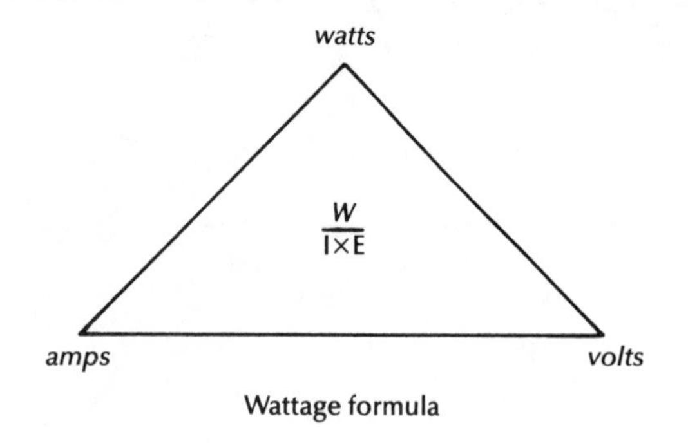


Wattage formula

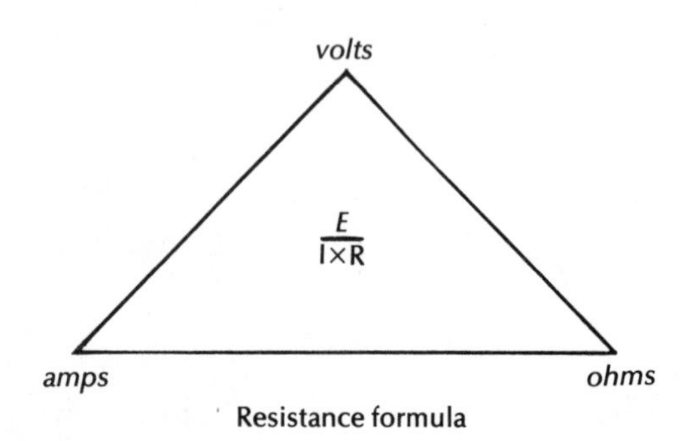


Resistance formula

Cover whichever unit you want to find. The uncovered part of the triangle will give you the formula for finding it. For example, if you need to know the amperage, cover the I in the wattage formula and divide volts into watts. If you want to find ohmage, cover the R in the resistance formula and you will know that volts must be divided by amps.

Note: This relationship can also be expressed as $W = I \times E$ and $E = I \times R$.

## HOW ELECTRICITY ORIGINATES

Power-generating plants throughout the world are continuously producing electricity in huge generators. The electricity is transmitted through enormous cables to various distribution stations. From each distribution station, it travels through overhead or underground cables that are tapped by smaller branch service cables

Where electricity comes from and how it reaches your home

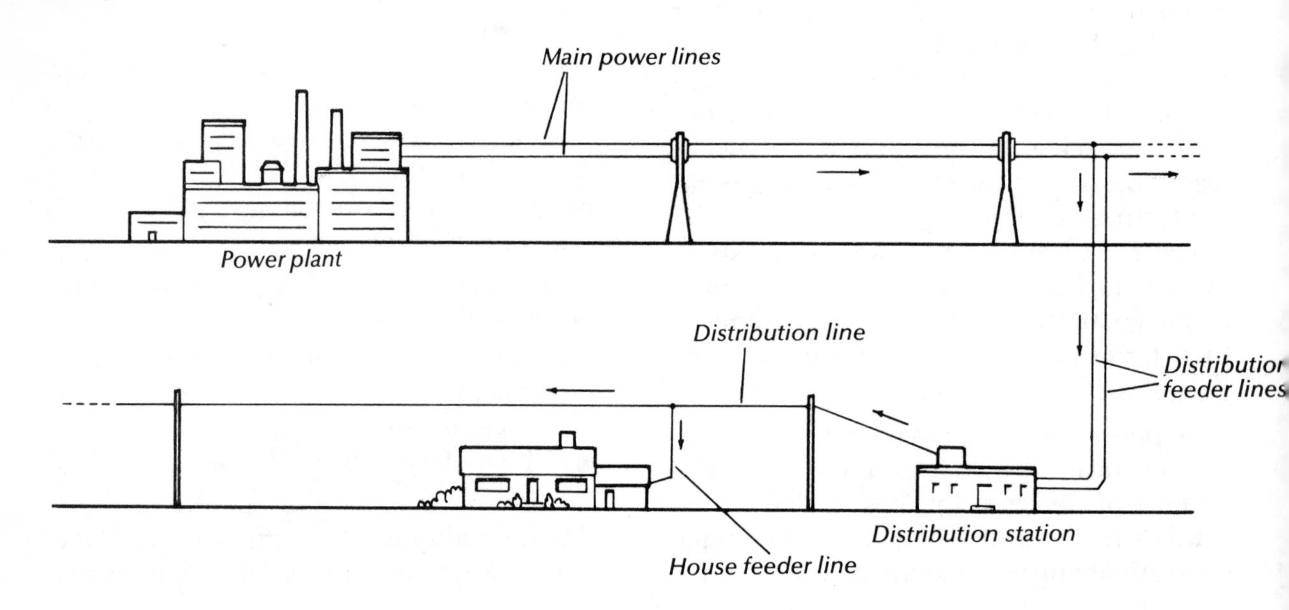

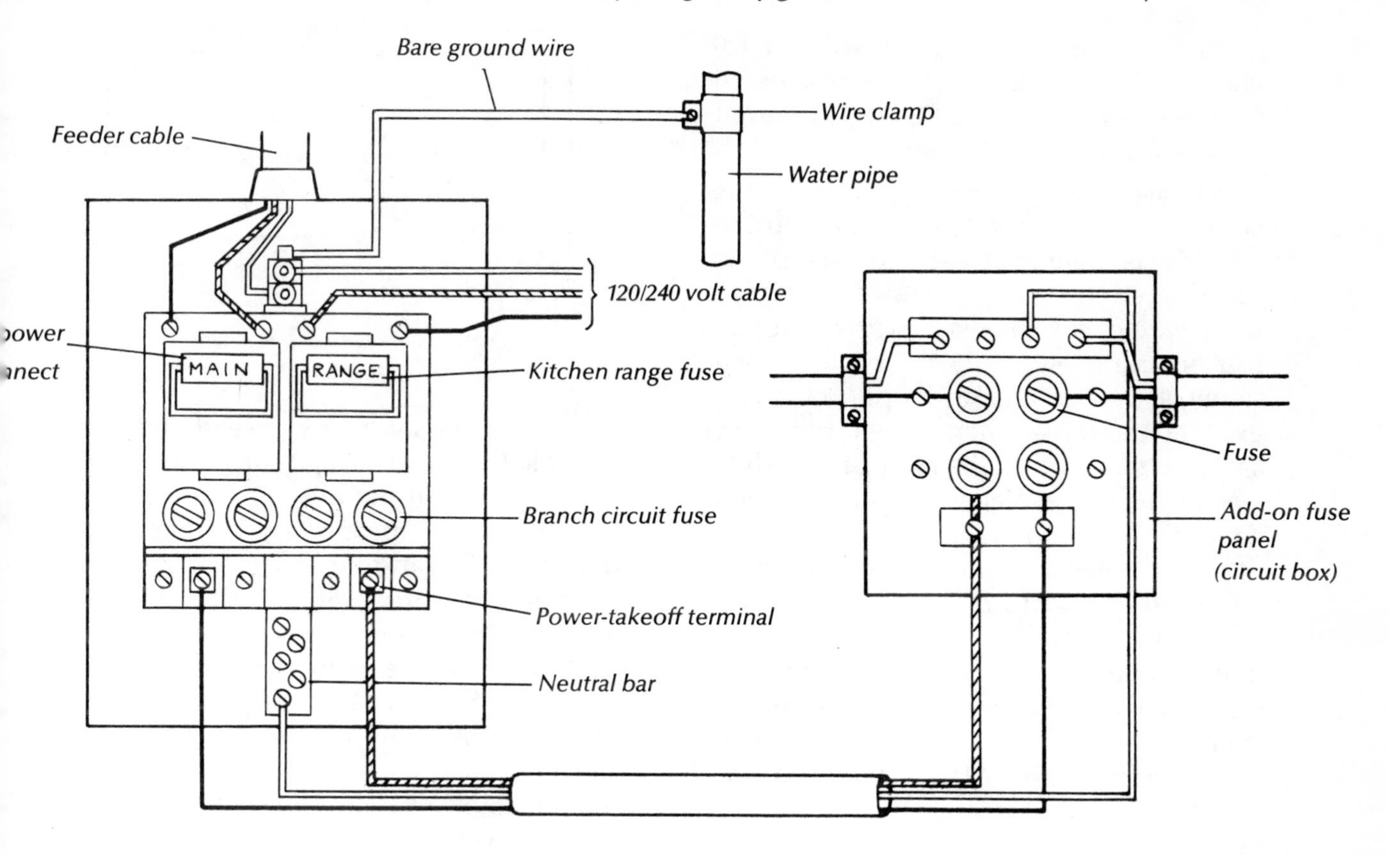


*Three-wire service provides 120 and 240 volts*

leading to each building in the area they serve. Local power companies maintain all of the equipment outside your home, plus your electric meter. Any other electrical equipment in your house is your responsibility.

The cable leading into your house has either two or three wires. If the house was electrified before approximately 1945, it probably has a two-wire circuit that brings 115 volts into the main power panel. The white wire is known as the grounded wire. The black wire is the hot line that brings the electricity to the distribution box.

Two-wire circuits provide 30 amps, or 3,600 watts, and are not adequate for most modern families. Larger refrigerators, washers, and TV sets, as well as most of today's appliances, came into being after World War II. You can ask your local power company if it will provide you with more power to increase the electricity entering your house.

Three-wire circuits serve most of the houses built since World War II. A three-wire circuit provides both 120 and 240 volts of electricity, which is enough to service both large and small appliances. Three wires enter your main power panel. The white wire is the grounded neutral wire; normally it is non-current-carrying. The two black (or one black and one red) "live"' wires each carry 120 volts of electricity. If you connect either of the hot lines with the neutral wire to an outlet, you will get 120 volts. If the two hot lines are connected together to service an appliance, the result is twice 120 volts, or 240 volts.

**Warning:** The neutral must be continuous—in a two-wire circuit you can attach a separate white wire to each of the receptacle's two silver screws, but in a three-wire circuit the white wire cannot be broken and must be attached to only *one* terminal, in a continuous loop. (In the one exception to this—when using cable—use a wire nut to connect the ends of the white wires plus a short piece of white wire, and attach the free end of the short wire to one of the receptacle's silver screws. *The connection must always be made on just one screw.*) If the neutral wire were not continuous and the receptacle was removed from the circuit (to repair it, for example), lamps and appliances plugged in past that point would receive 240 volts and burn out.

## House Voltages

Until recently, most houses received 110 or 220 volts of electricity. Today they receive 120 and 240 volts. However, in making electrical calculations use 115 and 230 volts, as suggested by the 1978 National Electrical Code, to give yourself a safety margin.

The modern three-wire circuit provides at least 60 amps, or a total of 14,400 watts of service. Newer houses are given 100 amps, offering 24,000 watts; 100 amps can meet the needs of a completely equipped residence, although central air conditioning or electric heating may demand another 50 or 100 amps.

## The Electric Meter

Whatever service enters your home, the electricity first goes through a kilowatt meter. Some meters are easy to read because their "dials" appear as digits in a row of tiny windows. A watt-hour is the number of watts consumed in one hour; a kilowatt-hour is 1,000 watt-hours. By reading the meter numbers from left to right, you immediately know how many kilowatt-hours were used from the moment the meter was turned on. Subtract the previous reading and you know how many kilowatt-hours were purchased since the last reading. Multiply the kilowatt-hours times the charge you are paying per kilowatt-hour to learn how much your electricity has cost you.

*Digital meters are read left to right*

*Read each dial according to the direction in which the pointer moves*

Utility companies charge between 5 and 10 cents per kilowatt-hour.

The typical kilowatt meter has four (or, infrequently, five) dials on its face, which are read from left to right. However, every other dial rotates *counterclockwise*. Each dial is numbered from 1 to 0; the pointer begins at 1 and rotates around to 0. You can identify the direction the pointer is moving by where the 1 is positioned on the dial face. If it is in the usual clock-face position, the pointer rotates clockwise. If the 1 is at 11 o'clock, the pointer moves counterclockwise.

When reading the dials, always take the number the pointer has just passed, *not* the one it is approaching. When you read a meter, write down the four numbers that each pointer has just passed.

# BRANCH CIRCUITS

From the main power disconnect, electricity enters your fuse or circuit breaker panel, where it is divided into branch circuits that carry it to each room in your house. Every branch circuit is composed of cables, wires, and outlet boxes, and is protected by a circuit interrupter in the panel.

## Three Types of Circuits

### *General-Purpose Circuits*

These are for lights and room outlets. General-purpose circuits are 120 volts and typically use #14 wire with a 15-amp fuse or circuit breaker to provide 1,800 watts. Some general-purpose circuits may be made of #12 wire and be protected by a 20-amp fuse or circuit breaker with 2,400 watts. Each general-purpose circuit is capable of serving 350–500 square feet of floor space. Do not put more than 10 outlets and/or lighting fixtures on any one circuit.

### *Special-Purpose Circuits*

Made up of #12 wire protected by a 20-amp fuse or circuit breaker, these provide 120 volts, 2,400 watts. They are used in workshops and kitchens, specifically to power small appliances. An average kitchen might have two special-purpose circuits with no more than 10 outlets on each one.

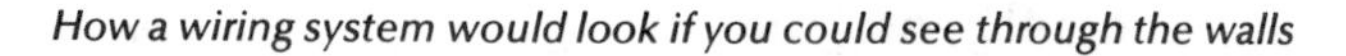

*How a wiring system would look if you could see through the walls*

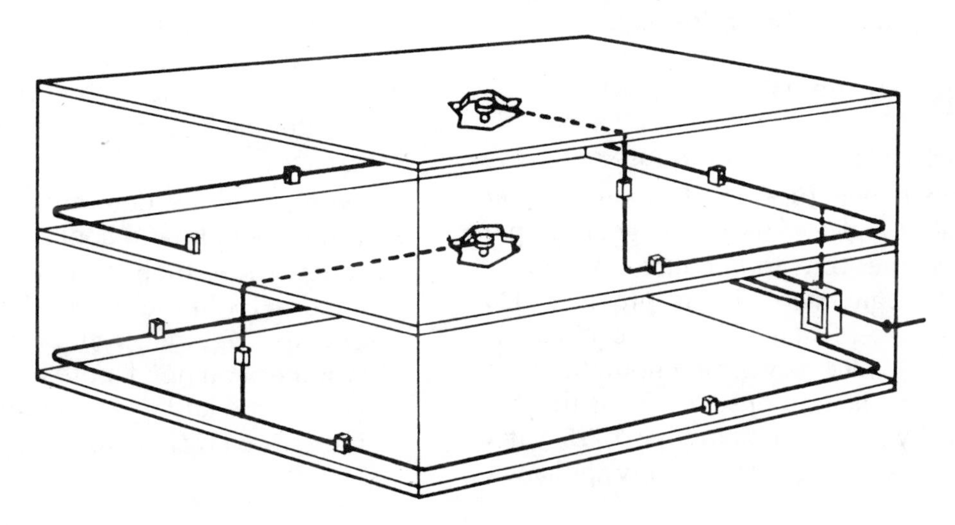

### *Major-Appliance Circuits*

These are three-wire circuits. The two hot lines together can provide 240 volts for a hot-water heater, electric range, or central air-conditioning unit. The size of the wires and the amperage of the circuit interrupters differ according to each appliance and its wattage rating; each circuit services only one appliance.

## WIRING LISTS

One of the first steps in upgrading your electrical capability is making a wiring list. With a complete wiring list you can determine which branch circuits feed each appliance and whether any circuit is overloaded. At that point, you can move appliances from one circuit to another, or you may decide to add another branch circuit. The list also tells you how much each appliance costs you to run, whether you are wasting electricity, and how many additional appliances your system will be able to support.

The chart pasted on the inside of the panel box door may already indicate each circuit location. But don't rely on the listing—it may not be up-to-date.

### *Making a wiring list*

1. Turn on all lights in your house.
2. Remove one fuse, or trip one circuit breaker, and note its amperage. Multiply amps times volts to calculate wattage.

   Although each circuit provides 120 volts (except major-appliance circuits), use 115 volts in your computations to provide a margin of safety. Thus the full capacity of a 15-amp circuit is 1,725 watts (15 amps × 115 volts). To allow appliance surge, the total load on that circuit should not exceed 80% of 1,725 watts, or 1,380 watts. A 115-volt, 20-amp circuit has 2,300 watts; all of its lights and appliances should draw no more than 1,840 watts. A 115-volt, 30-amp line provides 3,450 watts. The 80% margin allows a draw of 2,760 watts.
3. Note which lights in the house are not turned on. Plug a lamp into every outlet to find out whether the outlet is live. If it is not live, it is on the branch circuit you shut off.
4. Write down the wattage of every light and appliance on the circuit. Wattage is listed on most appliances. If the wattage is not listed, you will find the amps, volts, and/or ohms. Use the $W = I \times E$ or $E = I \times R$ formula (see page 44) to determine the wattage of each appliance.
5. Add up the wattages of every light and appliance on the circuit. Compare the total draw on the circuit with the 80%-capacity figure for that circuit. All of the electrical devices turned on simultaneously should not draw more than 80% of the wattage that the circuit is capable of delivering.
6. Replace or turn on the protective device. Remove the fuse or shut off the circuit breaker on a different circuit. Repeat steps 1–5 for the second branch circuit. Locate each circuit and compute its capacity.
7. When you have determined the wattage for every light and appliance, add them together. If their combined draw exceeds 80% of the power that all of your circuits together can provide, decide whether you need more power entering your home and whether you should add more branch circuits.

# CHAPTER 4 • Running New Circuits

The federal, state, and local governments, as well as fire insurance companies and the entire electrical industry, have spent untold time and money to make electricity and electrical equipment as safe as possible. The different governments have enacted rules and regulations designed to guarantee that all electrical systems are as hazard-free as they can be made. The National Fire Protection Association, backed by insurance companies among others, researches and publishes a precise book of electrical rules and standards entitled the National Electrical Code (NEC). It is the best bible to safe electricity that you could own. It is updated every other year, and if you cannot find it in a local store, you can order it directly from the National Fire Protection Association, 470 Atlantic Avenue, Boston, Massachusetts 02210 (617-482-8755 for current price).

*Underwriters' Laboratories, Inc.*

An independent organization, Underwriters' Laboratories, Inc., has the sole purpose of safety-testing electrical equipment that is manufactured for sale in the United States. The program is voluntary —manufacturers do not have to submit their products to UL for testing—but consumes regard UL so highly that they seldom purchase any electrical equipment that does not bear a UL lable. The label guarantees that the product has been exhaustively tested and is electrically safe *for its intended use.* So far as safety is concerned, any product that is not UL-listed might as well not exist.

## TOOLS

When you approach the project of adding branch circuits to your existing electrical system, you need only a few basic tools, most of which you probably already own.

*Screwdrivers*—You should have both Phillips-head and standard-slot screwdrivers of varying sizes. If you are buying screwdrivers for the first time, a typical seven-piece set provides about the variety of blade widths and thicknesses needed to tighten both large and small terminals and mounting screws.

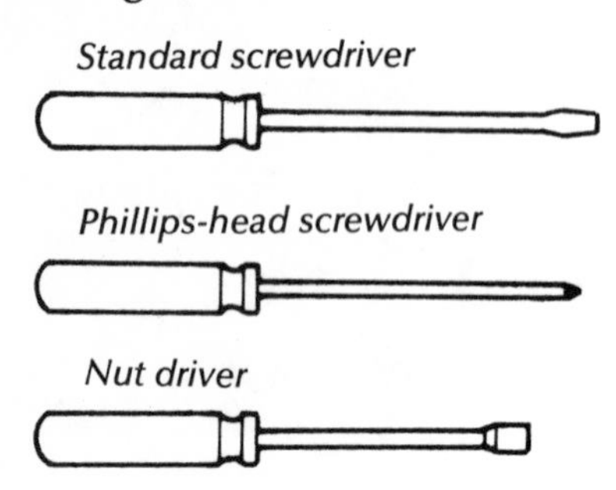


*Long-nosed pliers*—These are invaluable for holding small nuts and bolts and working with small wires, as well as bending cable wires to fit around screw terminals.

*Slip-joint pliers*—These are larger than the long-nosed pliers and are helpful when you are working with larger materials.

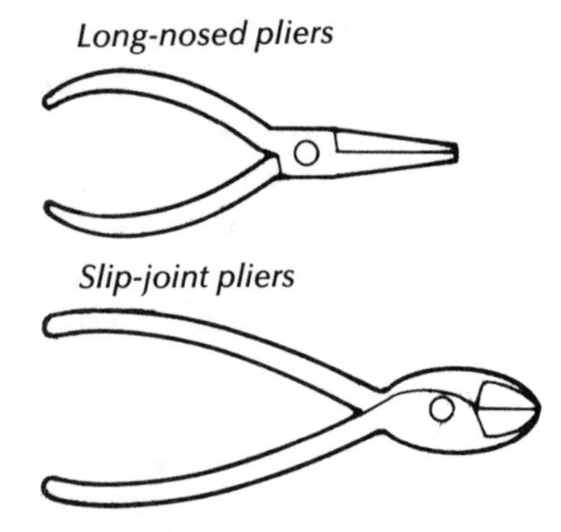


*Wire stripper*—Strippers have a series of different-size holes in their blades that are used to cut through the insulation around any gauge wire and remove it without damaging the wire.

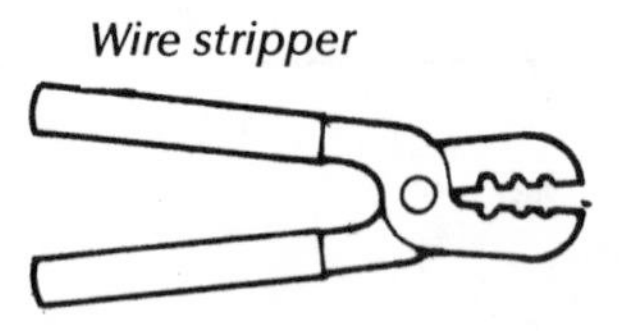


*Multipurpose tool*—This not only strips wire but also cuts it and tightens crimp-on splices and terminals over wire ends.

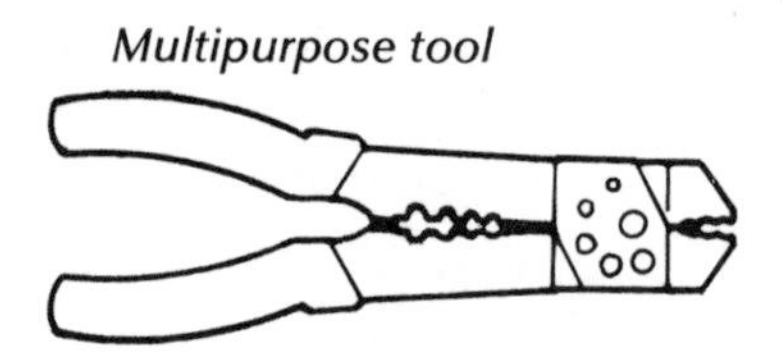


*Hacksaw*—You need this only if your local building code demands the use of armored cable.

*Utility saw*—This will cut laths, plaster, and wallboard, and is excellent for making square holes in walls to house electrical boxes.

## Neon Hot-Line Tester (continuity checker)

This costs about a dollar and consists of a tiny neon light, which can withstand 500+ volts of electricity without being damaged, attached to a pair of wires that end with small metal probes. To find out whether a wire or a piece of equipment is receiving electricity, touch the probes to different

Neon hot-line tester

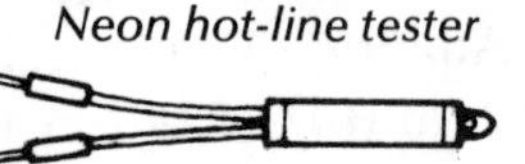

parts of the system to see if the light goes on.

*Testing receptacles*

1. Push one probe into the shorter slot of the receptacle. This should be connected to the black wire of the circuit cable.
2. Touch the other probe to the faceplate. If the faceplate is painted, use the mounting screw. If the screw is painted, take off the faceplate and touch the probe to the electrical box.
3. If the light goes on, the circuit is properly grounded. If the light does not go on, the grounding wire is improperly connected or broken.

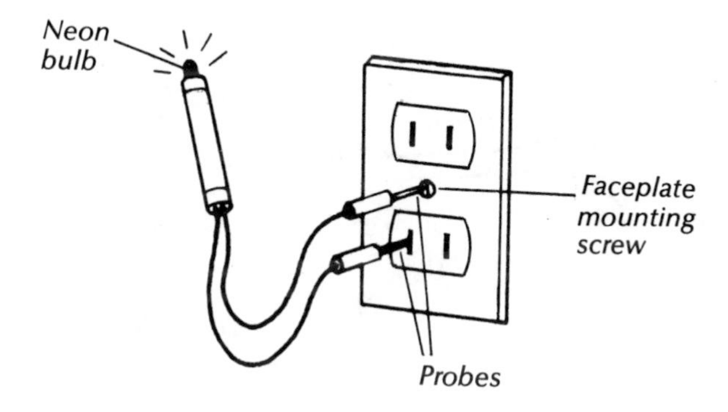


*Testing cartridge fuses*

You can test a cartridge fuse with a hot-line tester without removing the fuse from its holder. The procedure is described on page 13, step 3.

# EQUIPMENT

Aside from an assortment of wire nuts and a roll or two of plastic electrician's tape, the equipment needed to add a branch circuit to your home is primarily cable, electrical boxes, and wire staples, which are really just U-nails. U-nails can be driven in with a hammer or special staple gun.

## Electrical Boxes

Every switch, receptacle, and wire connection in every house system must be contained in an electrical box. The box may be rectangular, round, octagonal, or square. They all come with round knockouts die-stamped on their sides. Pounding the blade of a screwdriver against a knockout creates a hole that is the proper diameter to receive a cable clamp, which in turn grips the cable entering the box. Some boxes include a double cable clamp to facilitate holding the cables.

Many of the rectangular electrical boxes have removable sides to allow two or more boxes to be ganged together so you can install two or three switches or outlets next to each other.

There are screw-insertion holes all over the sides of the boxes; some boxes have metal flanges, soldered to their backs or one of their sides, with holes for nails, screws, and pointed prongs that can be driven into the wooden framing members in a wall or ceiling to secure the box. You can purchase metal supports to hold the boxes in drywall constructions, adjustable brackets that suspend them between ceiling joists, metal plates that are screwed to the open face of the box to seal the splices it contains.

Electrical boxes are the result of years of evolution and development, and no matter what situation you find yourself in, the chances are almost 100% that you can find a box to suit your purpose.

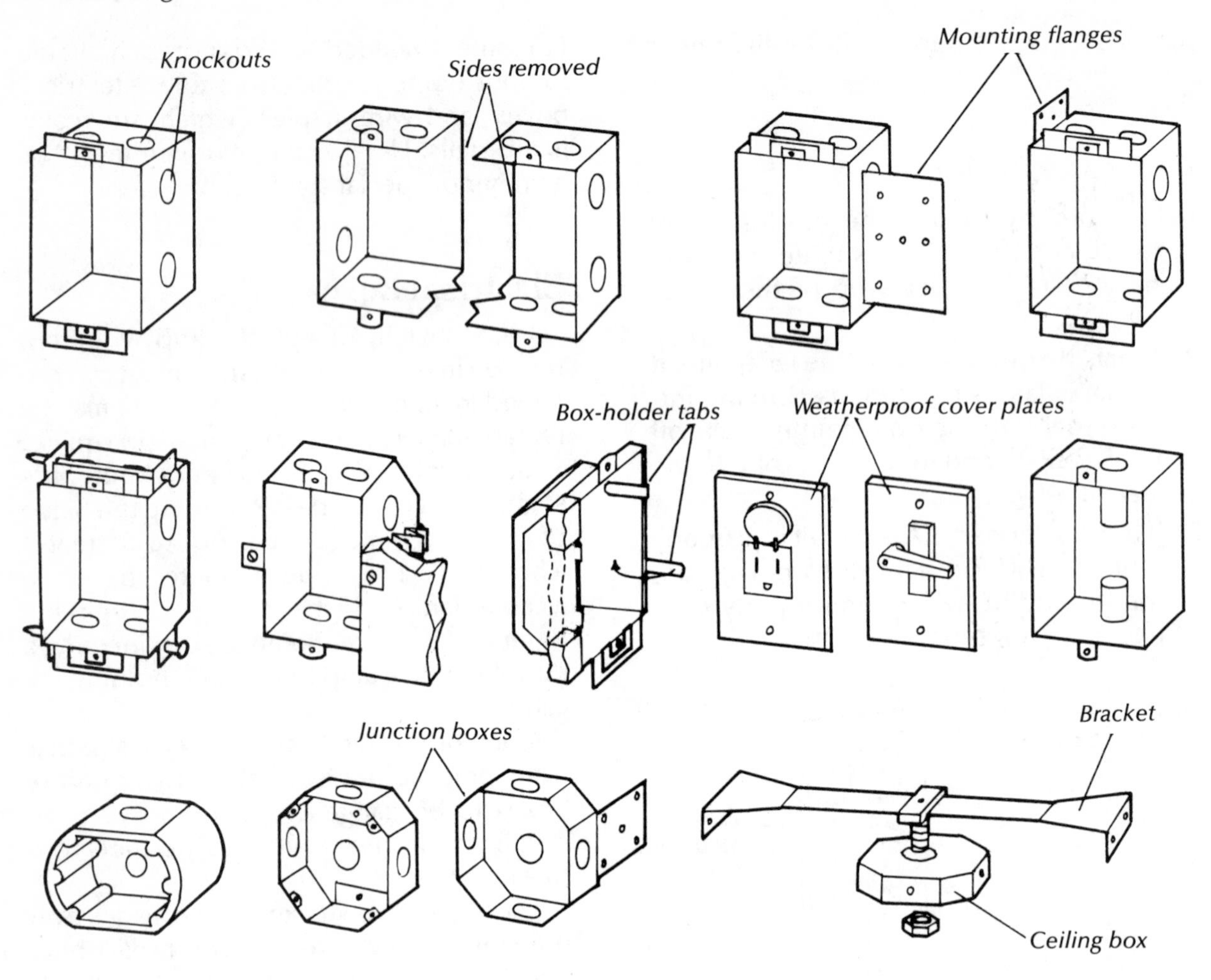


Electrical boxes

## Wire

The best electrical conductors are metal. As silver, copper, and aluminum offer the least resistance to the flow of electricity, most metals are not used at all. Silver is very expensive, aluminum wiring can be a fire hazard (any heat buildup in the wire tends to loosen its connections), so copper wire is the recommended conductor for use in a house electrical system.

All electrical wire is wrapped with an insulating cover of rubber or plastic and is given a number, based on its diameter, known as the wire gauge. Gauges are standardized by the American Wire Gauge system: the smaller the gauge, the larger the wire diameter. Gauge numbers go from 0 to 40. Gauges from 8 up to 0 are too rigid to bend readily, so they are made up of smaller-diameter stranded wires.

The wires used in house branch circuits are usually #10, #12, and #14. Number 16 and #18 are very thin wires used for door bells and intercoms.

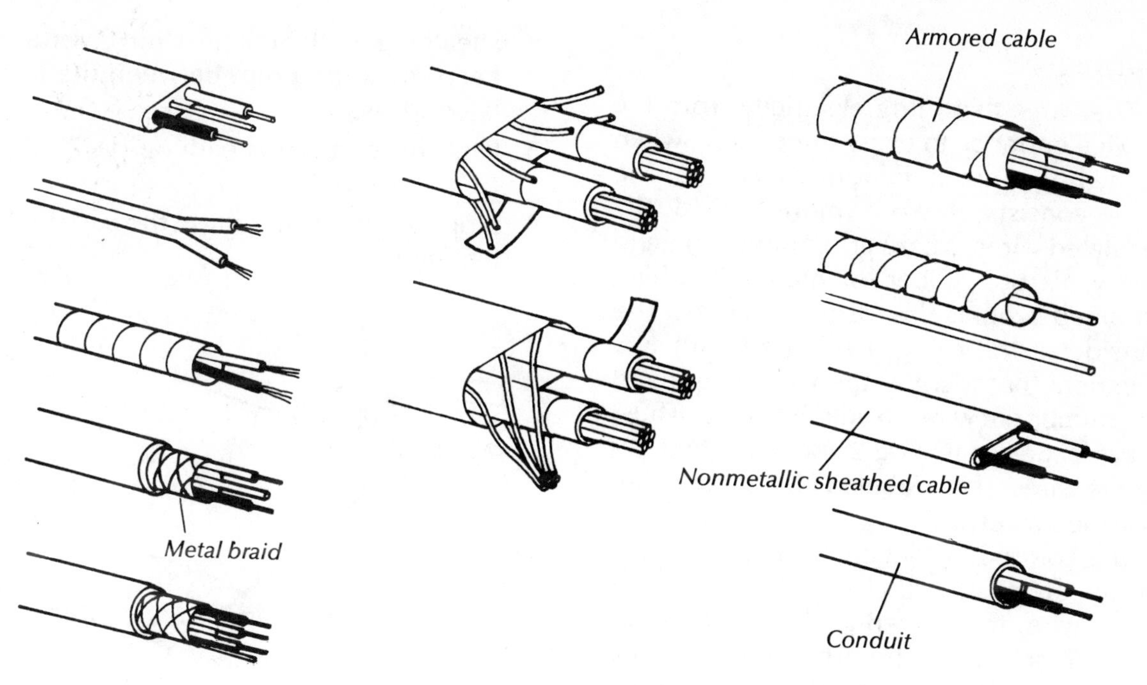


Wires, cables, and conduit

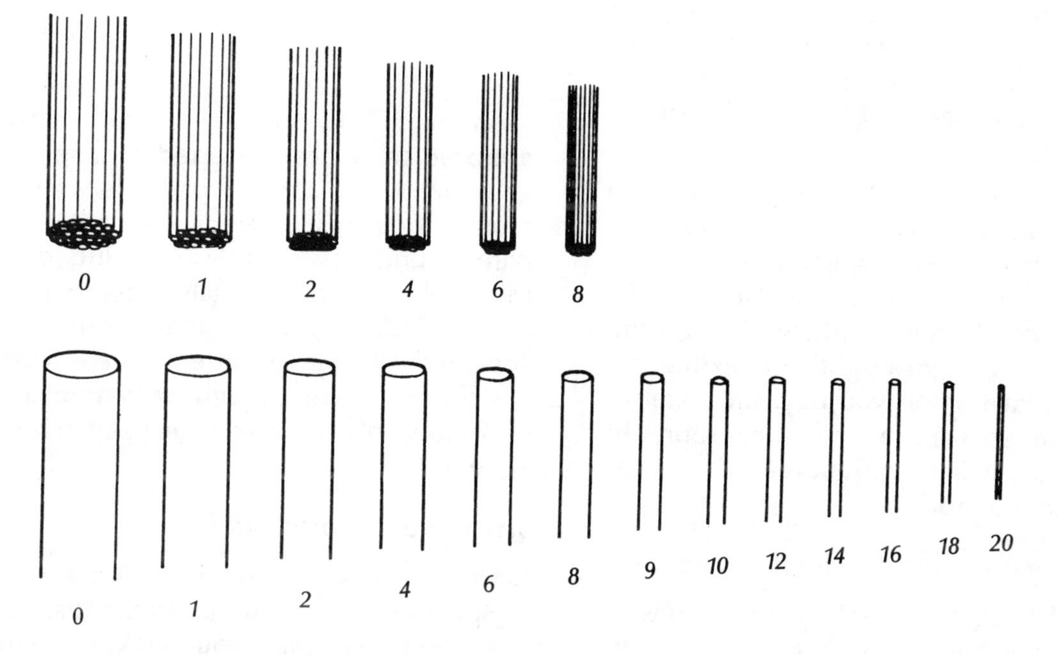


Wire gauges

## Cables

The wires that bring electricity from the service entrance to every outlet and switch in the house are in the form of a cable. The cable consists of two or more individually insulated wires, a bare grounding wire, and heavy plastic, rubber, or metal sheathing wrapped around the insulated wires. The outside of the cable is stamped with two numbers that identify the wire gauge and the number of wires inside the cable. Thus, a cable designated 12-2 contains two 12-gauge wires; if it is labeled 14-3, it has three 14-gauge wires.

The colored insulation around the wires inside any cable will always differentiate a white wire, to be used as the return wire, and a black, live wire. Three-wire cables also have a red live wire, and four-wire cable has a white, a black, a red, and a blue wire (usually live).

Two types of cable are currently used in residential electrical systems. Both can be purchased with different-gauge wires; the difference between them is their exterior sheathing.

*Nonmetallic sheathed cable*—Although this is sold in several different exterior sheathings that are manufactured for use indoors, outdoors, in dry or damp conditions (all of which is clearly marked on the sheath), it is all lightweight and flexible. No matter what version you buy, the exterior sheathing resists both moisture and fire. In most cases a bare grounding wire is included in the cable.

### Stripping nonmetallic sheathed cable

1. Cut the length of cable to be used with any wire-cutting tool (pliers, multipurpose tool, wire strippers).
2. Slice lengthwise down the center of the sheathing with a knife. Be careful to keep the knife point between the insulated wires.
3. Peel the severed sheathing back, away from the wires.
4. Strip each wire, starting at the base of the lengthwise slice.

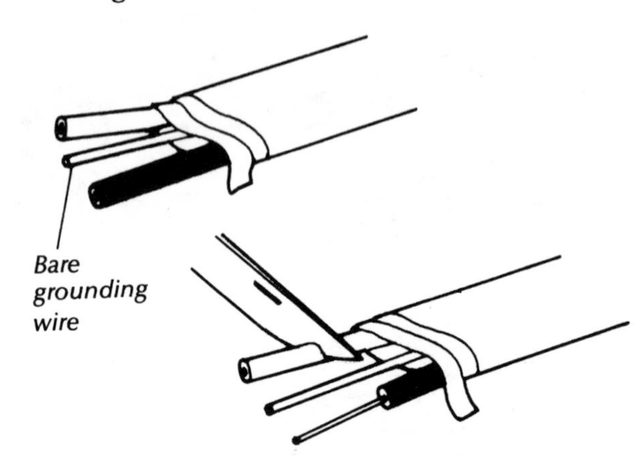


*Unwrap the sheathing (above) and bare the end of each cable wire (below).*

*Armored (BX) cable*—The wires inside armored cable are protected by a spiral of galvanized steel. Heavy paper is wrapped around the wires just under the steel, and a bare grounding wire is always included with the cable wires. The galvanized steel is much harder to bend than nonmetallic sheathed cable, so armored cable is harder to work with, even though many municipalities allow only BX cable to be used in home systems.

### Stripping armored cable

1. With a hacksaw, cut *diagonally* across the spiral of the sheath, not in the direction of the spiral. Angle your hacksaw so that it crisscrosses the spiral.
2. Saw through the sheath. Be careful not

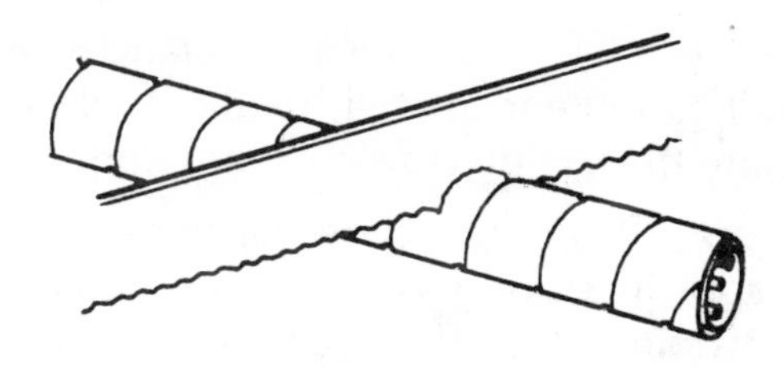

*Saw across the spiral*

to nick the insulation around the wires. Rotate the cable under your saw blade until you have cut completely around it.

3. Twist the severed end and pull it off the wires.

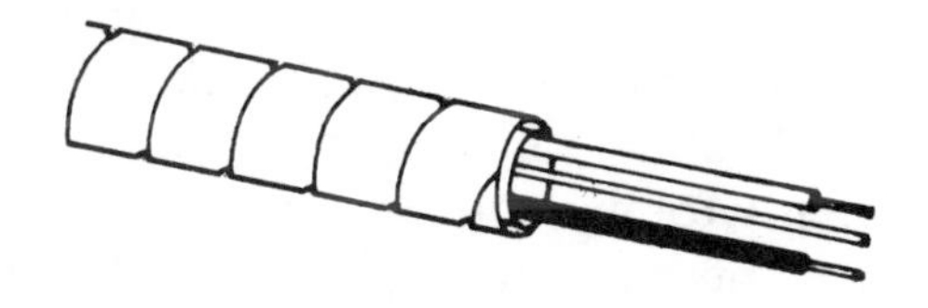

*Pull off the severed end*

4. The ragged metal edge of the sheath is sharp enough to cut through the wire insulation, so the National Electrical Code requires that a special fiber bushing be placed over the end of the armor. The bushing is split so it can fit around the wires and be pushed between the metal and the paper wrapping inside the armor.

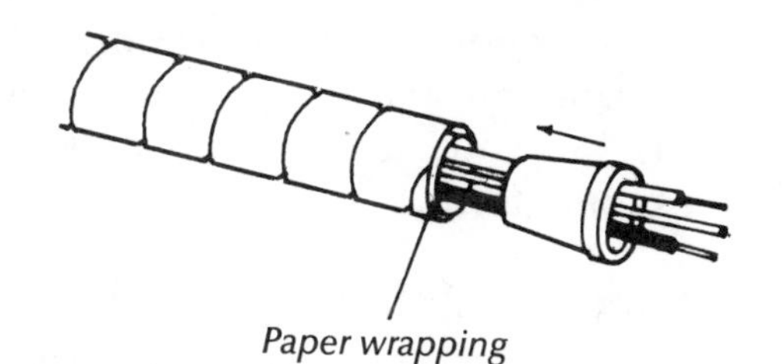


*Insert a protective fiber bushing*

5. Bend the bare grounding wire back over the edge of the bushing and wrap it tightly around the outside of the armor sheath.

# GROUNDED APPLIANCES AND CIRCUITS

There is a very real danger that water will in some way touch your electrical system. For example, if the insulation in a cable has worn with age and you plug in an appliance that is leaking current into the house system, you may get a shock when you turn on the appliance. If your hands are wet at the time, the shock could be lethal. Within the electrical industry there is great concern over the possibility of appliances causing fatal accidents, and a great many steps have been taken to reduce the hazards of electricity by grounding every circuit.

*Three-prong grounding plugs*—Many appliances are now manufactured with a three-prong grounding plug. The extra prong is round and will fit into the third hole in any of the new receptacles; its purpose is to dissipate any current leaking from the appliance directly into the ground, rather than through you or your house cir-

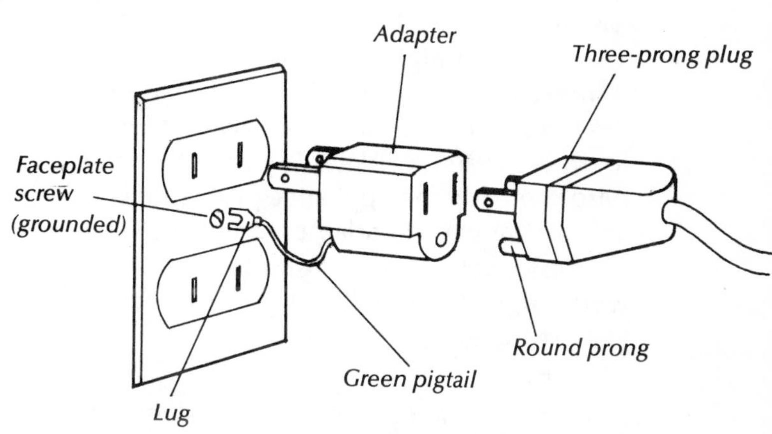


*Three-prong grounding plugs may require the installation of an adapter*

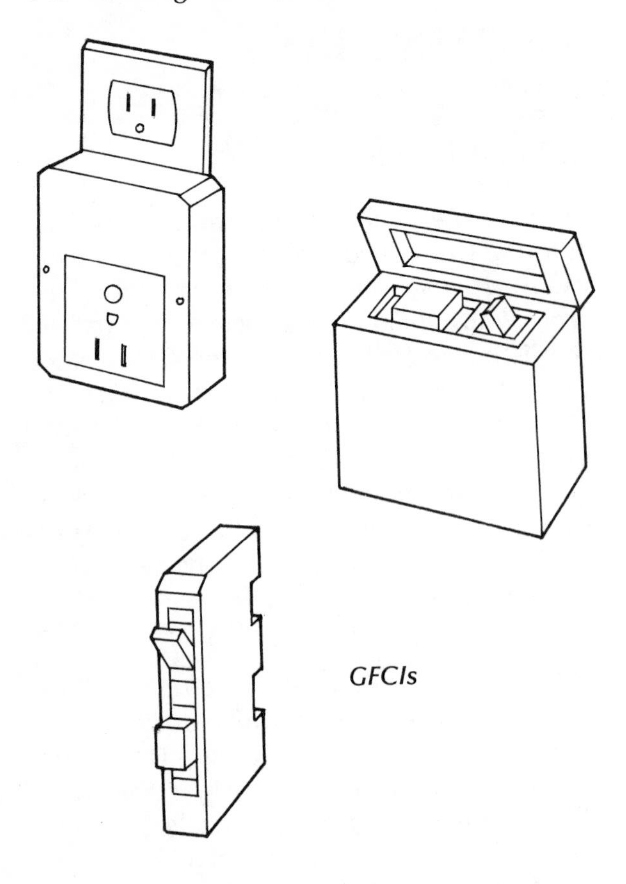

*GFCIs*

cuits. If the receptacle does not have a third hole, you can purchase an adapter plug, which is attached to the receptacle. The adapter has a pigtail wire with a crimp-on terminal, which should be placed around the (unpainted) mounting screw that holds the receptacle faceplate. If you cut off the third prong on the appliance plug, or if you fail to connect the adapter's pigtail to the faceplate screw, the grounding will not work and there will be no protection against receiving a shock from the appliance.

*Double-insulated appliances*—Manufacturers have begun making tools and many small appliances with an extra layer of insulation and plastic covers so that if part of the unit becomes live there is little chance of its causing a shock. These appliances are called "double-insulated," and they do not require the use of a three-prong plug.

*Ground-fault circuit interrupters*—While the appliance and tool manufacturers have been doing what they can to reduce the danger of using their products, the electrical industry has developed two types of ground-fault circuit interrupter (GFCI or GFI). One type is an adapter that is plugged into any grounded wall outlet, so that an appliance plugged into it is completely grounded. The second type of GFCI looks and performs like a circuit breaker and can replace any standard circuit breaker. But it is considerably more sensitive to anything that goes wrong in the branch circuit it protects or the appliances plugged into that circuit.

All GFCIs react to as little as 5 milliamperes of current leakage within 1/40 second, and break the circuit. Thus, if an appliance short-circuits (begins leaking current), the entire branch circuit will be turned off almost instantaneously.

GFCIs are recommended for use in every home by both the federal government and the electrical industry, particularly if the appliances being used are even remotely subjected to moisture of any sort. This means outdoor equipment, tools normally used in basements and garages, and most particularly any appliance used in kitchens, laundries, or bathrooms.

## Grounded Electrical Systems

Your share in the war against electrical hazard is to construct any branch circuits you add to your system so that they are completely grounded. When you are working with electrical cable you will always have at least three wires. The black wire is always the hot line and is always connected to other black wires. The white, return wire conducts electricity only when an appliance is turned on, and then it conducts that electricity away from the appliance, back to the fuse or circuit breaker panel. With the one exception involving switches (see page 35), the white wire is always connected to white wires. The third, bare wire provided in most electrical cables is known as the grounding wire. It is always attached to whatever electrical box it enters, by a special grounding clip or a screw terminal.

A properly grounded electrical circuit is one in which every outlet, switch, and junction box, every inch of cable, and all exposed metal parts are connected by the third, bare wire in the cable. When a three-prong appliance cord is plugged into such a circuit, it continues the grounding system all the way through the appliance. At the other end of the circuit, the service panel is grounded by a wire that leads from the panel to a cold-water pipe that goes immediately into ground. The white wires from each branch circuit are also connected to a grounding bar in the panel box.

# PLANNING A BRANCH CIRCUIT

Look at your existing circuit breaker or fuse box and determine whether there are any unused socket positions. If there are, you can bring a new branch circuit into each position; if there are no positions available, you can connect an add-on panel box to the main service panel. Most main entrance panels have a screw terminal located between the two left-side fuses and a second terminal between the two right-side fuses. These are meant to be used as power-

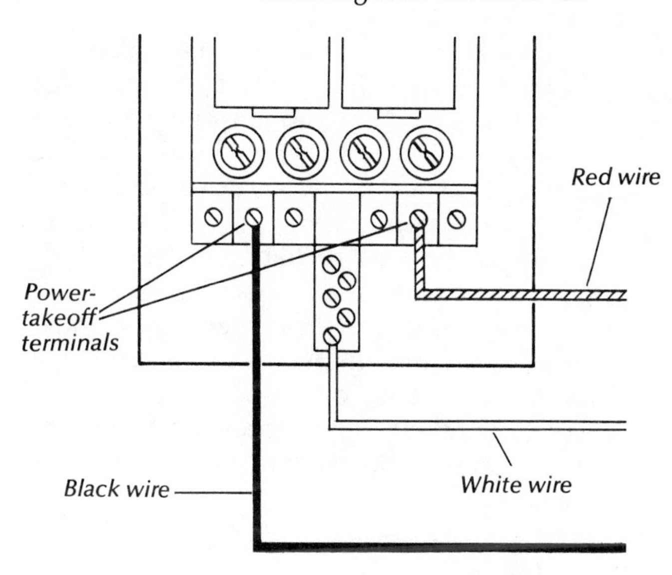


*Use the power-takeoff terminals to wire an add-on panel box*

takeoff terminals, so if you find that they are not in use, you can add on a panel box containing four fuse or circuit breaker positions to handle new circuits (see page 71).

*Check with your utility company*—If you intend to add branch circuits, you will be drawing more electricity into your house —*if* the power lines from your local utility are large enough to provide it. Call your local utility and ask if the capacity for giving you more current is there. Utility companies are delighted to sell you more electricity, but the equipment they are using to serve you right now may have to be changed, or may be already operating at capacity. So ask first. There is no point in going to all the work of running new circuits if the utility cannot service them.

*Talk to your local building department* —Every municipality has specific building codes governing electrical work. Some locales flatly do not permit homeowners to do their own electrical work, but most areas will let you do everything except connect your new branch circuits to the service panel; the final hookup must be done by a licensed electrician. The local building code will also give you a precise guide as to how each connection must be made in order to pass the building-department inspection. While the inspection may seem an annoying step in major electrical work, it is an assurance that you will not be living with a potentially dangerous electrical system that might someday cause a fire or injure someone in your family, or a future tenant.

The local code will also tell you what type and gauge of cable must be used, and which electrical equipment is considered safe. In most cases, #12 wire is recommended throughout a house, although #14 is sometimes allowed for 15-amp general-purpose circuits.

## Plan Each Circuit

Decide where you want each circuit to run, and how it will get there from the service panel. You can attach the cables to the sides of joists in the basement, but then they must go up through your house inside the walls to reach each outlet and switch you want to have on the circuit. Try to keep the circuits for lighting separate from appliance circuits; if two circuits serve the same room, put the light fixtures on one circuit and the outlets on the other. Then, if the lighting circuit gets turned off, you can plug a lamp into an outlet.

General-service circuits should be designed to serve no more than 500 square feet of floor space with a 120-volt, 20-amp circuit, or 375 square feet of floor space with a 120-volt, 15-amp circuit. In either case, outlets should be spaced between 7 and 12 feet apart along the walls of each room, but there should never be more than 10 outlets on any one circuit.

Appliance circuits should be 120-volt, 20-amp circuits, and allow for at least one outlet for every four feet of counter space in the kitchen. The size of a major-appliance circuit depends on the single major appliance (water heater, furnace, clothes dryer) that it serves; #10 wire is normally used with 240 volts and a 30-amp circuit interrupter.

## Measuring and Buying Cable

Measure the distance from the panel box to the farthest outlet on each circuit, following the path the cable must take. Be liberal about your calculations and then purchase the appropriate amount of cable in 25-, 50-, or 100-foot rolls. As prescribed by the NEC, residential wiring must be no smaller than #14 with a 15-amp circuit interrupter. Many areas demand at least #12 wire, which can be fused with 20 amps. However, the wire gauge is also affected by the length of the branch circuit. If you use a #14 wire with a 15-amp fuse, it will provide 1,800 watts for as much as 30 feet. If you need to go farther than that with a 120-volt, 15-amp circuit, you must go up to a #12 wire. The chart below gives you the wire gauge and what wattage you can get from it for different maximum circuit lengths.

| WIRE GAUGE | CIRCUIT WATTAGE | | | |
|---|---|---|---|---|
| | 2,400 | 1,800 | 1,200 | 600 |
| #14 | — | 30′ | 45′ | 90′ |
| #12 | 36′ | 47′ | 71′ | 142′ |
| #10 | 57′ | 75′ | 113′ | 225′ |
| # 8 | 90′ | 120′ | 180′ | 360′ |
| # 6 | 143′ | 191′ | 280′ | 573′ |

# RUNNING A BRANCH CIRCUIT

Before pulling any cable through the walls, ceilings, and floors, mark where you want each of the outlets and switches to be on the walls. Outlet boxes are normally placed 12″–15″ above the floor; whenever you can nail or screw one to a wall stud, it will be more secure. Switches are usually 36″–42″ above the floor.

Having located all of the outlets and switches, you can go through your house and cut out the holes for them. Basically, the cable will be cut into lengths that reach from box to box with no splices in between.

## Positioning Electrical Boxes

### *In lath and plaster*

1. Mark where you want the box on the wall.
2. Gouge out the plaster until you uncover one complete lath.

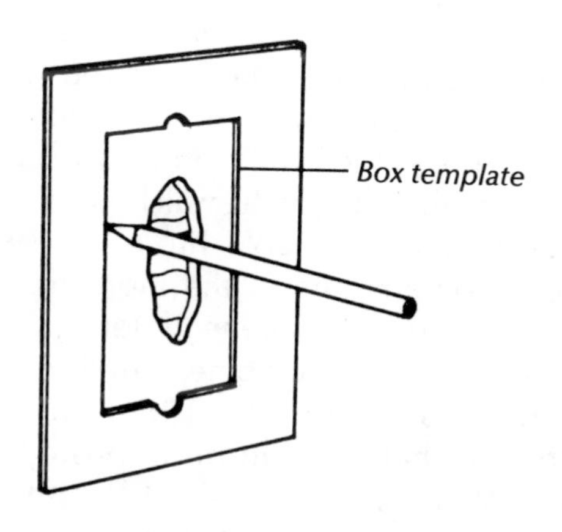


*Trace around the box or use a template to mark its position on the wall*

3. Center the box over the lath; trace its outline on the wall.
4. Cut away all of the plaster inside your outline.
5. Saw out only as much of the lath as needed to fit the box in the wall. What is left of the top and bottom of the lath will be used to hold the mounting screws in the box. If the hole is next to a stud, you can nail or screw the box to it.

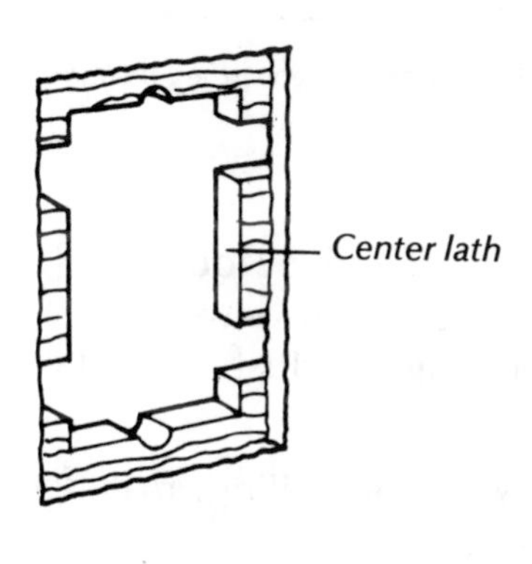


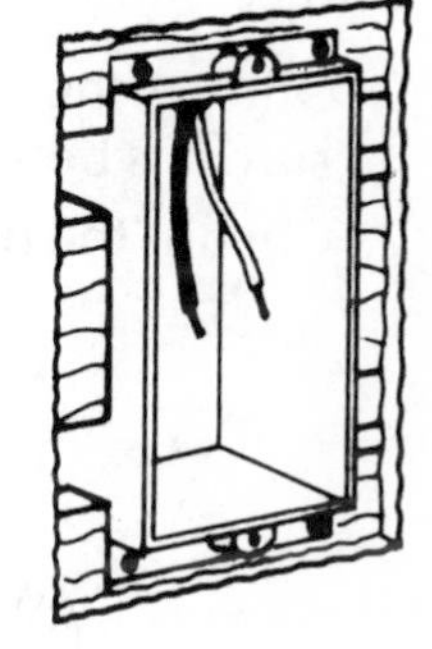

*Saw out the center lath and half of the top and bottom laths so the box will fit in the wall*

### *In wallboard*

1. Outline the box on the wall.
2. Cut out the outline with a utility saw. Be sure the hole is just large enough for the box to fit snugly.
3. If the hole is next to a stud, nail or screw the box to the wood. Otherwise, hold the box in place by sliding flat metal hangers into the hole on either side of the box. Tabs on the hangers fold over the box to hold it.

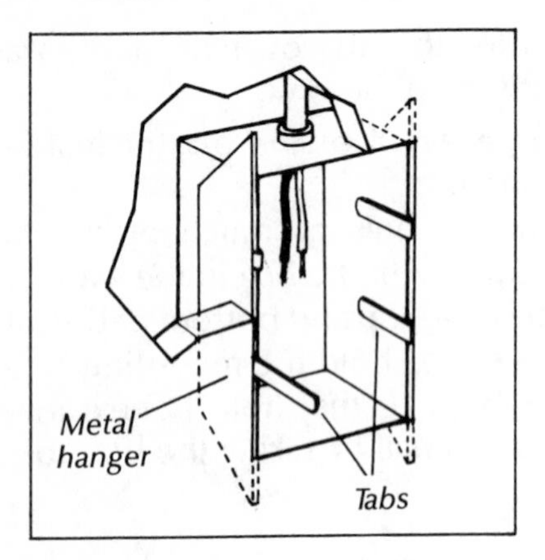


Boxes can be held by tabbed metal hangers; the tabs fold over the edges of the box

### In ceilings

Ceiling boxes are round or octagonal, and they must be positioned between the joists so that their edge is flush with the surface of the ceiling.

1. Outline the box on the ceiling, making sure it is between the joists.
2. Cut out the outline with a utility saw.

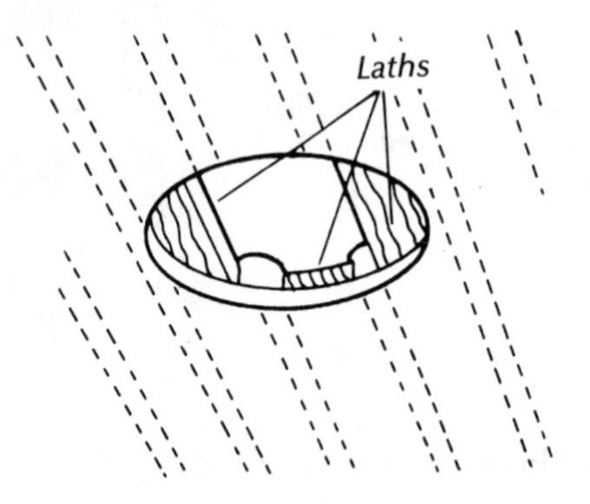


Saw out the outline of the ceiling box

3. An adjustable hanger that fits tightly between the joists can be inserted through the hole and wedged against the joists, then nailed or screwed in place. To do this you may have to enlarge the hole beyond the size of the box. If the ceiling is lath and plaster, you can use a short hanger that merely rests on the lath. Center the adjustable screw bolt on the hanger so that it hangs down through the center of the box hole.

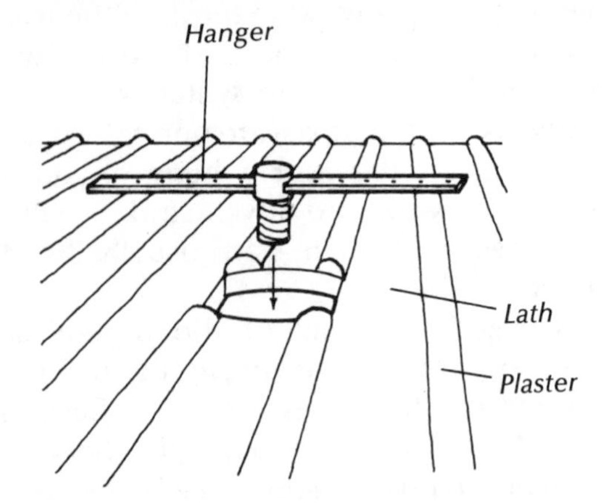


A metal hanger can rest on a lath and plaster ceiling

4. Screw the electrical box to the threaded pipe on the hanger.
5. Tighten the locknut to the threaded hanger pipe to secure the box in place.

## Running Cable

Unless you are going only from the basement to the first floor, it is usually easiest to run cable from the farthest outlet to the service panel. You must drill holes through the 2″×4″ bottom plate of each wall, then through the floor, the subflooring, and the top plate of the wall beneath. Then poke the cable down through the holes to the floor of the wall below. Without question, fishing cable through the walls of your home is one of the more frustrating chores you will ever encounter.

Two pieces of equipment are indispensable: an 18″ extension bit for your electric drill, so that you can bore through walls, wall plates, and beams; and a set of fish tapes. These are spring wires with hooks at each end and are used singly or together to fish wire through walls and ceilings.

### Fishing drywall construction

1. Drill holes through the studs and joists with the extension bit.

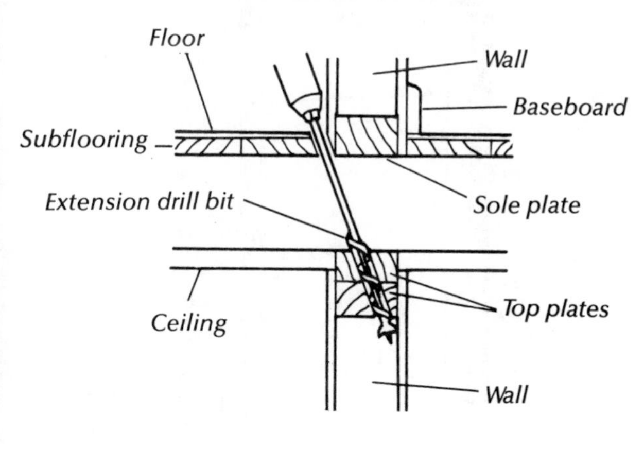


*Inside a wall and ceiling*

2. Feed one tape through the electrical box hole past the holes in the studs or joists.
3. Feed a second tape through the next box hole and up to the first tape.
4. Wriggle the tapes until their hooked ends connect.

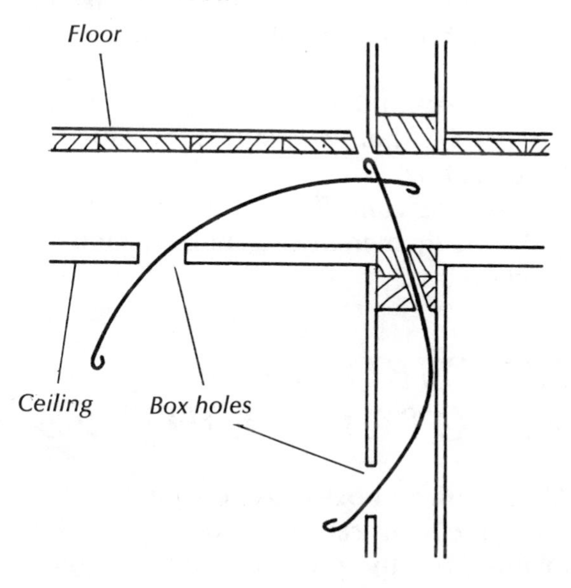


*Wriggle the fish tapes until their ends lock together*

5. Attach the end of your cable to one of the tapes.
6. Pull the unattached tape until the cable emerges from the other hole.

You can save yourself considerable time and effort by going through the top of ceilings or under open floors and into an attic or crawl space, then running your cable in open spaces above or below the room you are wiring.

### Fishing through lath and plaster

The lath and plaster construction found in older houses does not leave very much space between the walls. You can try fishing cable in that limited space, but when you are going across the studs, the best system is to channel the plaster.

1. Gouge a trench in the plaster deep enough to hold the wire.
2. At the point where the cable reaches the wall box, cut a hole through the plaster.
3. Fish the cable to the outlet hole and connect it to the box.

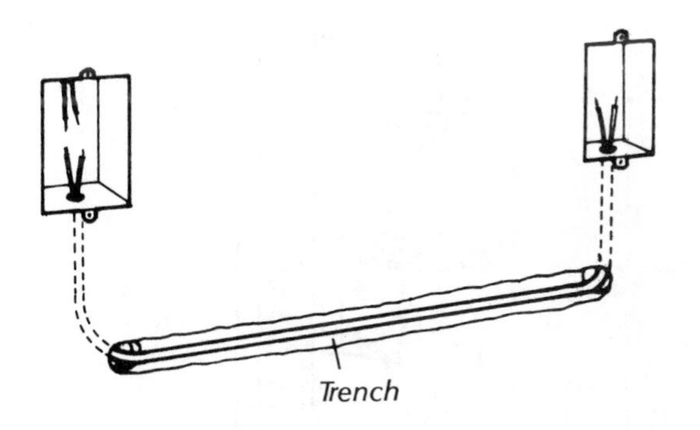


*Gouge a trench in the plaster when going across studs, then fish the cable to the boxes*

### Fishing around doors and windows

1. Bring the cable to the door frame.
2. Remove the door trim.

3. Run the cable up the space between the doorjamb and its supporting studs.
4. Drill through any uprights and headers you encounter.
5. Draw the cable up one side of the door, across the top of its header, and down the other side to the next outlet hole.

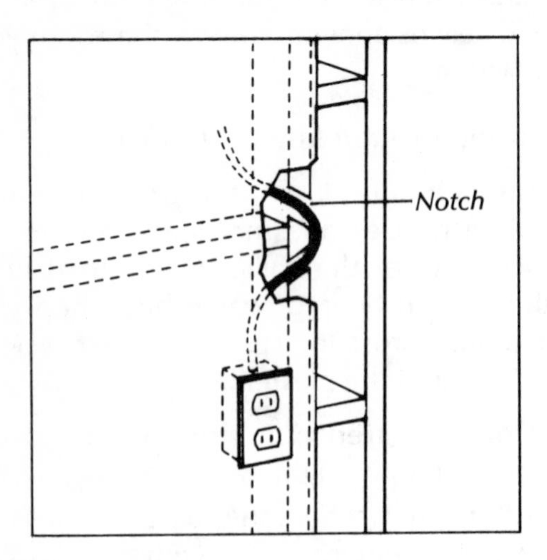


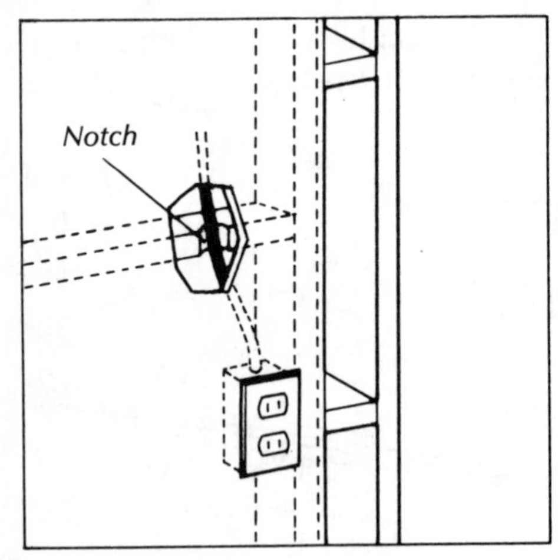


*You may have to notch frames and studs*

### When the Wire Is Too Short

If you have miscalculated the length of cable, you have only three options:

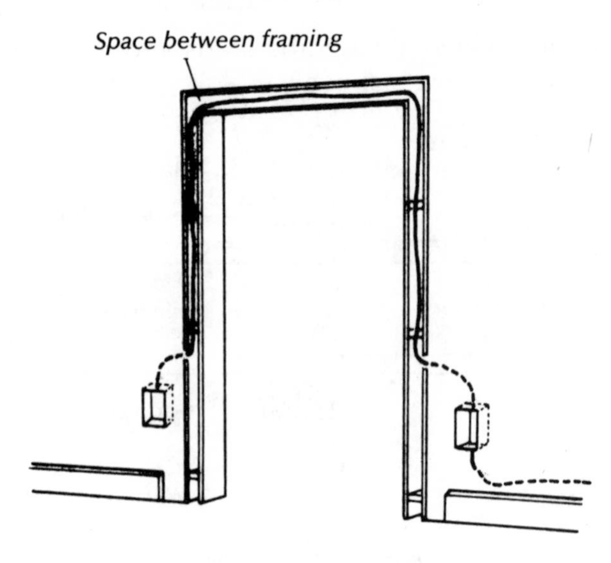


*Running cable around a door frame*

- Splice a longer cable to the end of the short piece and pull the short piece until the longer cable emerges from the wall. Then remove the short cable.
- Cut a hole for a junction box and splice more cable to the short wire inside the box. Bury the box in the wall and continue your cable run.
- Add an outlet at the point where the wire ends, then continue the cable run from the box.

What you *cannot* do is splice the wires and leave them in the wall unprotected by a junction box.

## CONNECTING ELECTRICAL BOXES

All electrical boxes are connected to a branch circuit cable in about the same manner, but the procedure differs slightly depending on whether you are using armored or nonmetallic sheathed cable.

## *Hooking up a box with armored (BX) cable*

1. Remove a knockout in the side of the box for each cable that will enter or leave the box.
2. Strip 8″–12″ of the armor sheath from the cable end. Cut the sheath at an angle across the spiral with a hacksaw. Remove the paper insulation around the wires.
3. Straighten all ragged edges in the armor with pliers. Insert fiber bushing over the edge to protect the wires. (If a building inspector cannot see any bushing when he inspects your work, he will reject it.)
4. Wrap the bare grounding wire back around the armor.
5. Strip ¾″ of insulation from the wires.
6. Insert the cable through the cable clamp in the box. Tighten the screw against the armor.

If you are using a box that has no cable clamp, put a cable connector in the knockout hole. The connector has a threaded end, which goes inside the box to accept a flanged nut. Tighten the nut against the inside of the box by hammering a screwdriver blade against the flanges on the nut. Insert the cable through the connector. Tighten the screw in the outside of the connector until it holds the cable in place.

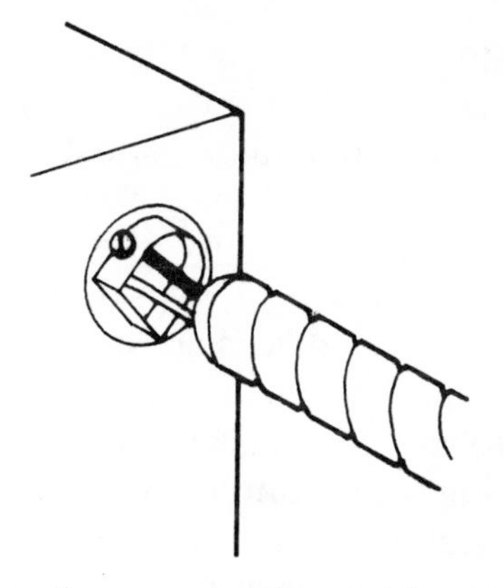

*Insert the armored cable and tighten the screw against it*

## *Connecting nonmetallic cable to an electrical box*

1. Strip 8″–12″ of sheath from the wires.
2. Electrical boxes used with nonmetallic

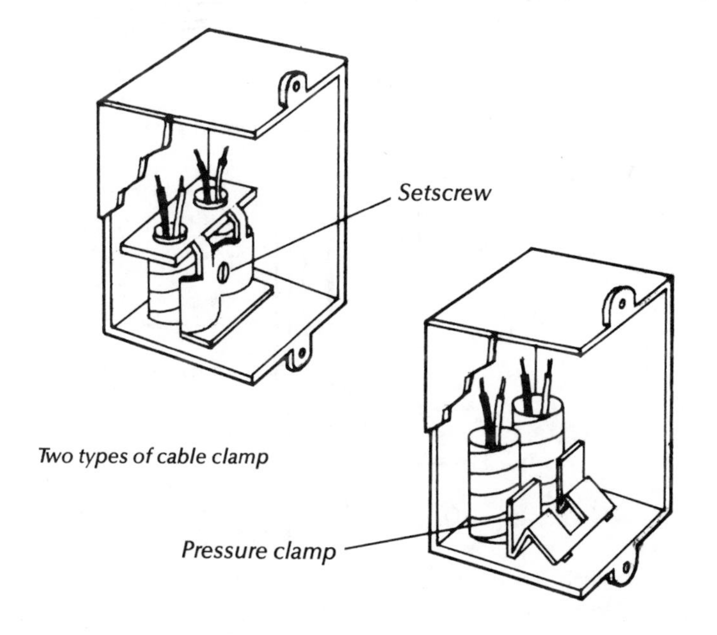


*Two types of cable clamp*

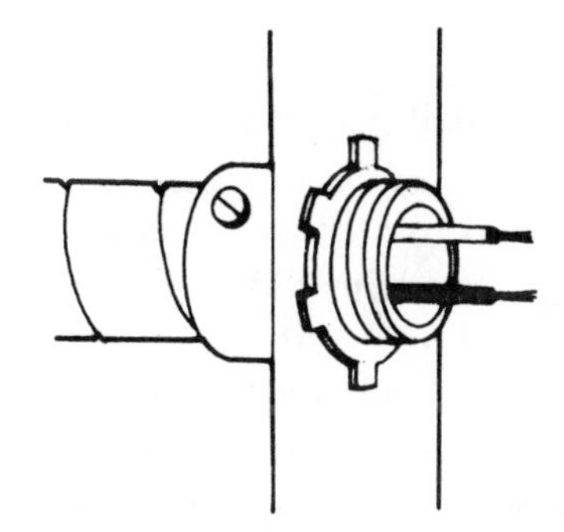

*Tighten the screw against the armor, the nut against the inside of the box*

sheathed cable have a special cable clamp that locks the cable in place with a screw. Nonmetallic sheathed cable clamps consist of a threaded clamp that is tightened against the cable with screws that bring a yoke down over the sheath. Connectors are held to the box with a flanged nut.

3. There are four accepted ways of connecting the grounding wire in nonmetallic sheathed cable:

A. Connect the bare wire to a screw inserted in any of the tapholes in the box.

B. Insert the bare wire in a special grounding clip that fits over the side of the box.

C. The grounding wire can be bent back and held with a grounding strap attached to the box with screws.

D. Squeeze the grounding wire under the cable clamp before you tighten it to the electrical box.

## Two-Wire and Three-Wire Cables

Two-wire cable will serve most or all of your branch circuit needs, but in a kitchen or a workshop, where you are likely to use heavy-duty electrical equipment such as refrigerators, dishwashers, or power tools, give yourself the options available with a three-wire cable. When connecting three-wire cable to an outlet box, you always have the choice of using the white line and one or both hot lines. With a single black wire, the outlet will receive 120 volts. The extra black or red wire is then spliced to the third line of the cable leaving the box. With both

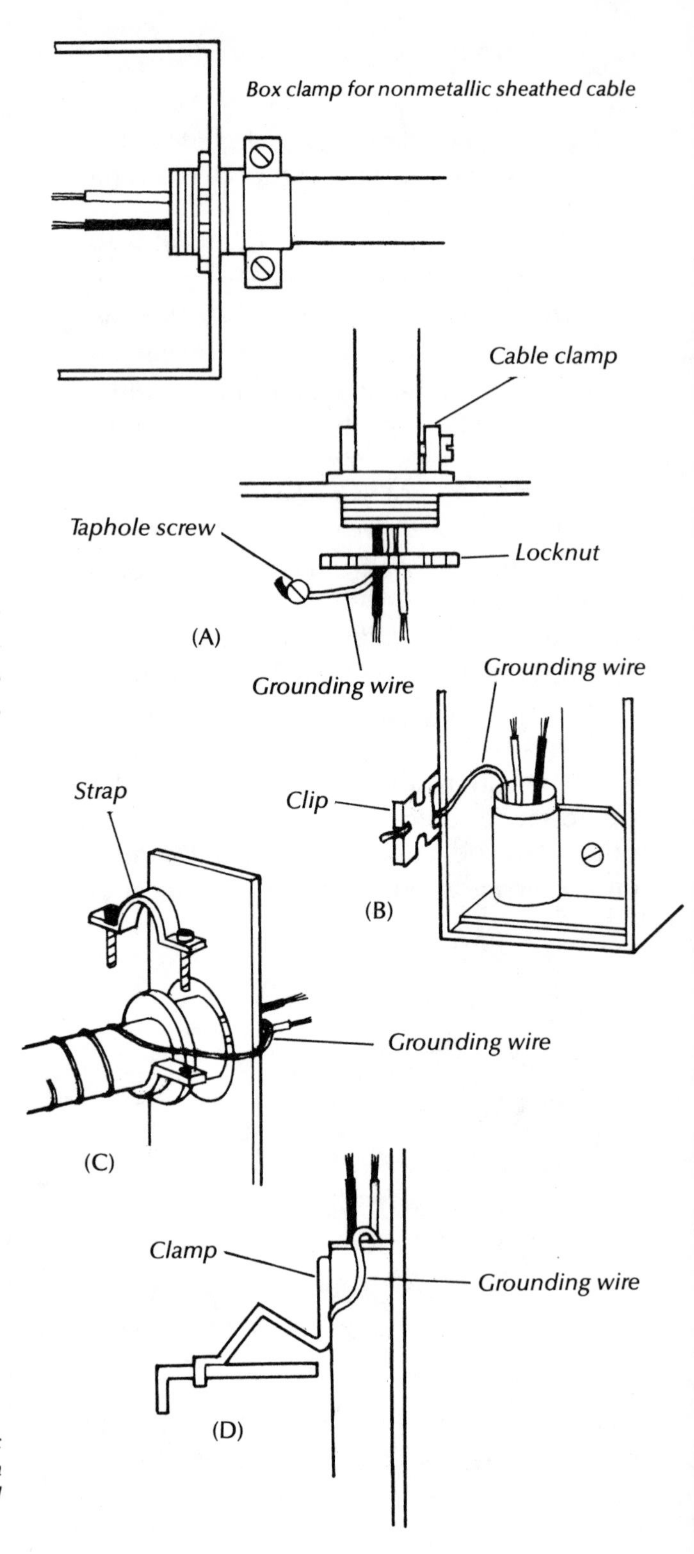


*Four ways to connect the grounding wire in nonmetallic sheathed cable: (A) Connect it to a screw inserted in a taphole. (B) Insert it in a grounding clip. (C) Bend it back and secure it with a strap. (D) Squeeze it under the cable clamp*

hot lines in the cable attached, the outlet will receive 240 volts.

Another time where three-wire and even four-wire cable comes into play is when you are connecting switches to light fixtures and/or outlets. There are a dozen standard wiring arrangements used to connect various combinations of lights, switches, and outlets, but no matter how they are connected, they and the cables that connect them must also be grounded to their boxes.

*Wall switch controlling one light at the end of the cable run (Fig. 1)*

The black wire from the switch is connected to the black wire from the fixture. The white wire from the switch is connected to the black wire in the cable and painted black at both ends or wrapped in black tape to signify that it is being used as a hot line. The white wire from the fixture is connected to the white wire in the cable.

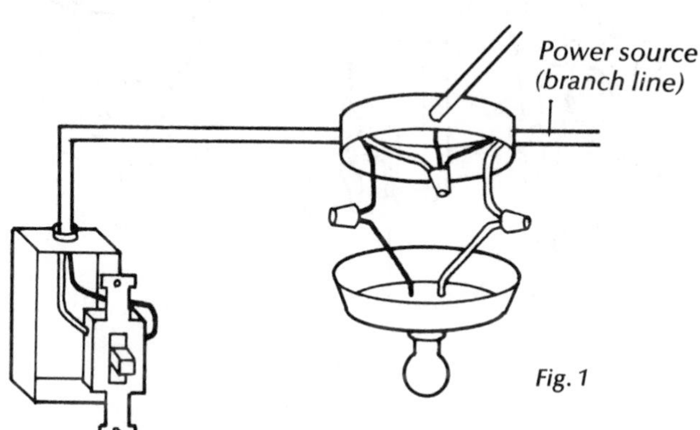

Fig. 1

*Wall switch controlling one light in the middle of a cable run (Fig. 2)*

In the middle of a run there are at least two cables entering the fixture box. All of the white wires from the cables will be connected to the white wire from the fixture. The black wires from the cables are attached to the white wire from the switch. The black wire from the switch is connected to the black wire from the fixture. The white wire from the switch must be painted black at both of its ends.

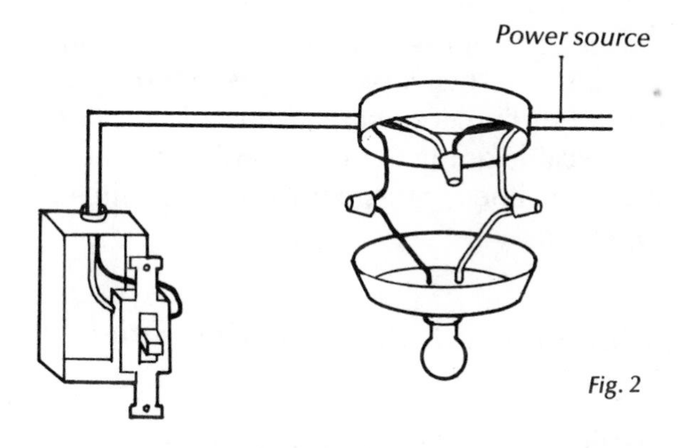

Fig. 2

*Two fixtures controlled by separate switches (Fig. 3)*

If one of the switches is on one of the fixtures and the other switch is in the wall, you need a piece of three-wire cable to connect the wall switch with its light. The fixture with its own switch simply has its black and white wires connected to the black and white wires in the cable, whether it is in the middle or at the end of the cable run. The light with the wall switch has its

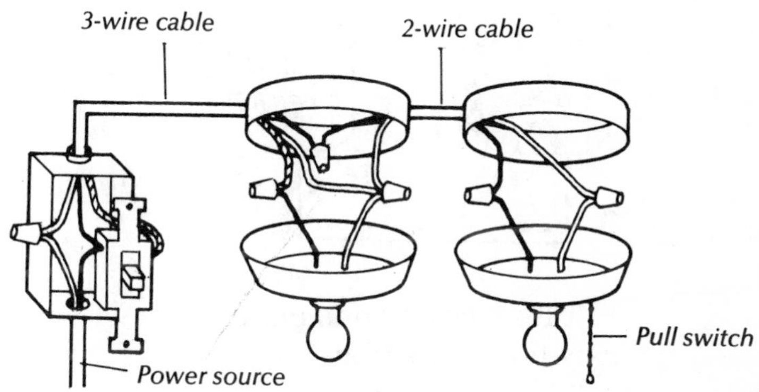

Fig. 3

power source at the switch box. The power cable enters the switch box, where its black line connects to the switch. Its white line connects to the white wire of a three-wire cable running out of the switch box to the light. Both the hot lines in the three-wire cable also connect to the switch. At the fixture, one of the three-wire hot lines connects to the black wire in the fixture. The other hot line connects with the black line in the cable going to the second light. All of the white lines in the fixture box are spliced together.

### *Adding an outlet to an existing outlet (Fig. 4)*

Paired receptacles have two brass and two silver screw terminals. The power source (the branch line) is connected to either the top or bottom set of terminals (black wire to brass screw; white wire to silver screw) in the old box. The cable to your add-on outlet box is connected to the second set of terminals and to either the top or bottom set of terminals in the new receptacle. Be sure to ground both outlets to their boxes.

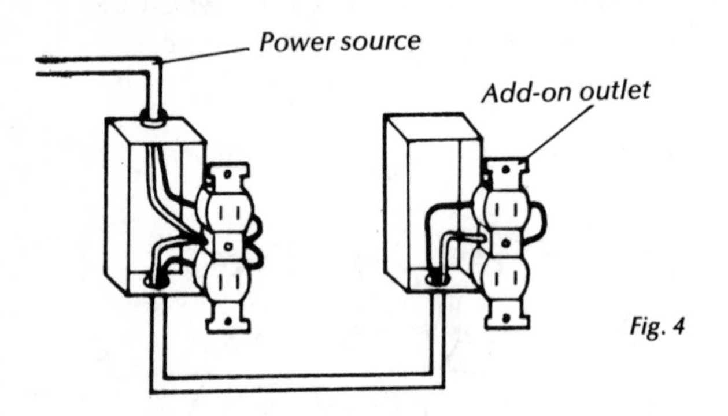


Fig. 4

### *Adding a switch and outlet to an existing fixture (Fig. 5)*

The outlet must operate independently of the wall switch so that it can be used even when the light is turned off. To achieve this you need a length of three-wire cable between the fixture and the switch. The white wire in the three-wire cable is connected to the white wires of the power cable and the fixture. One hot line in the three-wire cable is connected to the black wire in the power cable; the other hot line is attached to the black wire in the fixture. The two hot lines in the three-wire cable are both connected to the switch, along with the black wire of a two-wire cable leading to the outlet. The white wires from the outlet cable and the three-wire cable are joined in the switch box. The outlet cable's white wire is connected to the silver terminal on the receptacle, its black wire to the brass terminal.

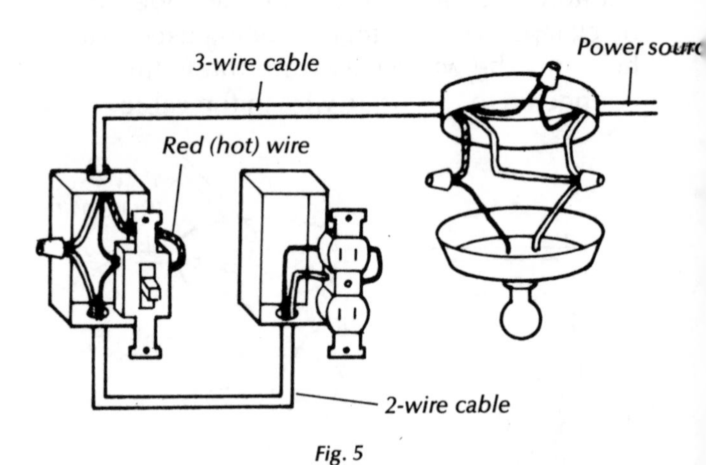


Fig. 5

### *Adding a fixture or outlet to a junction box (Fig. 6)*

The junction box will have several cables entering it. Run a separate two-wire cable from the box to a fixture or outlet box and connect it to the unit. At the junction box, the white wire in the new cable is connected to the existing white wires. The black wire is connected to the black wires.

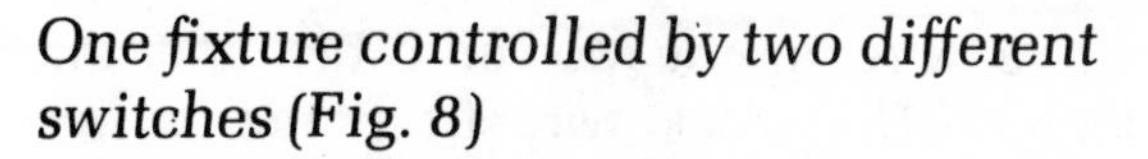

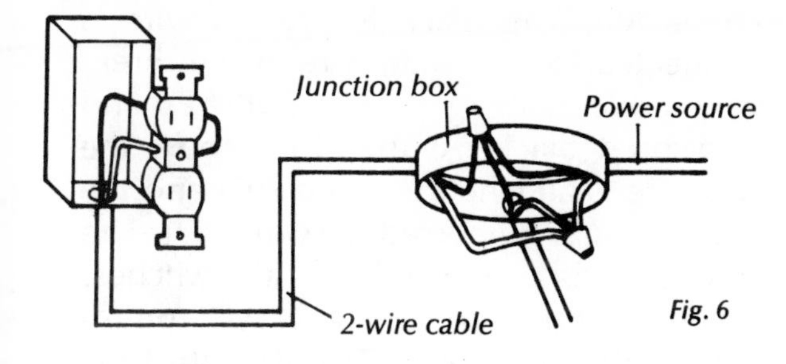

Fig. 6

### *Adding a switch and outlet in the same box to a fixture (Fig. 7)*

You need two rectangular electrical boxes. Remove the left side of one box and the right side of the other, then hook the boxes together and anchor them in the wall. Run a three-wire cable from the double box to the fixture. At the fixture, the white wires all go together. One hot line of the three-wire cable is connected to the black wire from the fixture. The other hot line is spliced to the hot line in the power cable. In the double box, the hot line connected to the black fixture wire is hooked to the switch and the white wire is attached to the silver screw terminal on the outlet. The second hot line has ½″ of insulation stripped from its middle, and is looped around the open terminal on the switch. The wire is then extended to the outlet and connected to one of its brass screw terminals.

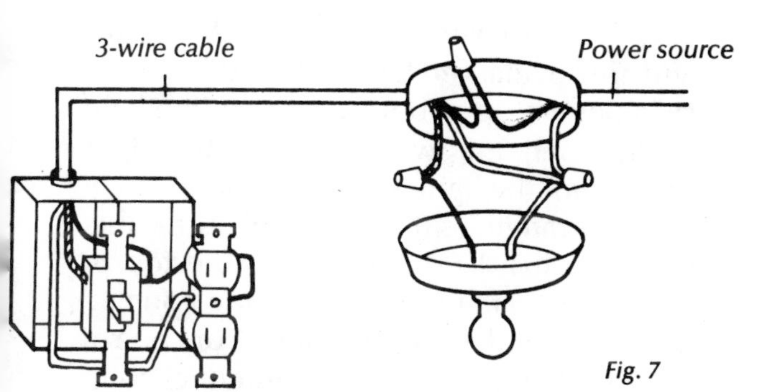

Fig. 7

### *One fixture controlled by two different switches (Fig. 8)*

You can control a light with either of two switches, but they must be interconnected with a three-wire cable. One of the switches (A) is attached to the fixture with a two-wire cable. The black wire is connected to the switch and the black wire in the power cable. The white wire is connected to the black wire from the fixture and one of the two hot lines in the three-wire cable. This wire must be painted black at both ends. The other hot line in the three-wire cable and the white wire are connected to switch (A). The three-wire cable enters the second switch box (B), and all three of its wires are connected to screw terminals on the switch.

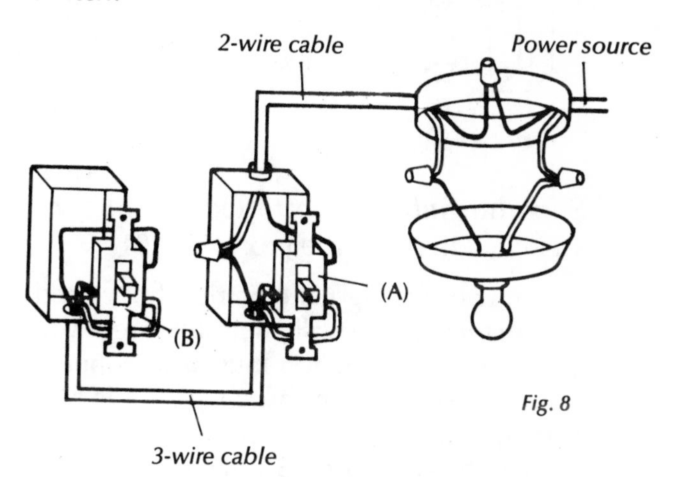

Fig. 8

### *Controlling a fixture between two switches (Fig. 9)*

Both switches are connected to the fixture with three-wire cables. The three wires in each switch box are all connected to the switch. At the fixture box the two red wires from the switches are connected. The two white wires from the three-wire cables are also connected, but separately from the white wire in the power cable and the fixture. The black wire from the power cable is connected to the black wire in one of the switches (A). The black wire from the fixture is connected to the black wire from the other switch (B).

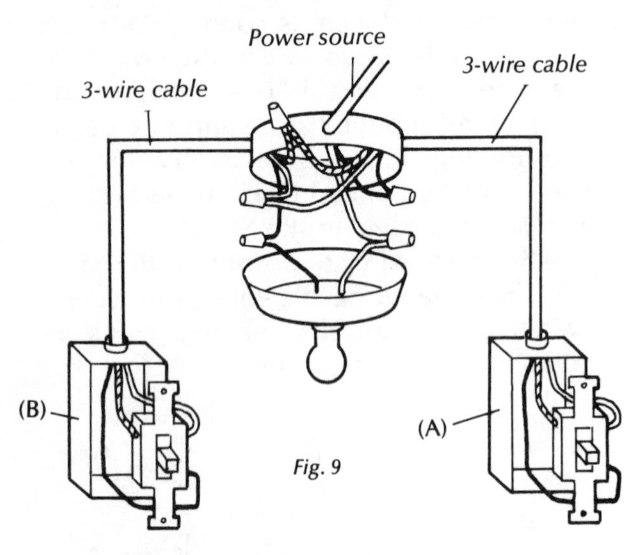

Fig. 9

## *Separate fixtures controlled by different switches in the same box (Fig. 10)*

Join two rectangular electrical boxes and secure them in the wall, then run three-wire cable to fixture (A). You must also connect the two fixtures with three-wire cable. Assuming the power source is a two-wire cable entering fixture (B), make these connections: Fixture (B)'s black wire is spliced to the red wire in the three-wire cable connected to fixture (A). The black wires from the three-wire cable and the two-wire power cable are connected. The white wires from the power cable, fixture (B), and the three-wire cable are spliced together. In fixture (A), the white wire from the three-wire cable is spliced to the white fixture (A) wire. The black wires from both three-wire cables are connected. The two red wires are connected. The black fixture (A) wire is connected to the white wire in the three-wire cable leading to the switches and is painted black. At the switch box, the red wire is attached to switch (C) and the white wire is connected to switch (D). The black wire is connected to both switches. Given this hookup, switch (C) will control fixture (B) and switch (D) will control fixture (A).

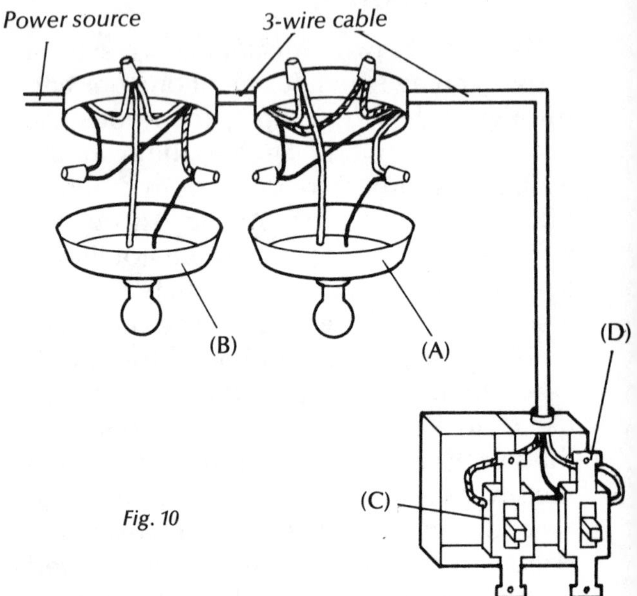

Fig. 10

## *Fixture controlled by three-way switches; outlet connected to fixture is always hot (Fig. 11)*

Three-way switches have a double throw and are not marked On and Off. When both switches are up or both are down, the light is off. When one switch is up and the other one down, the light is on. This is the kind of switch arrangement that you might have to control a hall light from either the top or bottom of a stairway. To make the connections you need three-wire cable from one of the switches to the fixture and either four-wire cable or 2 two-wire cables between the switches.

The power source is two-wire cable leading into switch (A). Four-wire cable leads from switch (A) to switch (B). In switch (A), the two white wires are connected together. The black wires from each cable are connected together with black jumper wire, which leads to a screw terminal on the switch. The red and blue wires from the four-wire cable are connected to the switch. At switch (B) the four-wire cable enters one end of the box and a three-wire cable leaves the box and goes to the fixture box. The white wires in the two cables are connected, as are the black wires. The red and blue wires from the four-wire cable are connected to the switch, as is the red wire from the three-wire cable.

At the fixture box, the red wire from the three-wire cable leading out of switch (B) is connected to the black wire from the fixture. The black wire from the three-wire cable is spliced to the black wire from the two-wire cable leading to the outlet. The white wires from the fixture, the three-wire cable, and the two-wire outlet cable are connected together. The two-wire cable leads to the outlet box; its black and white wires are connected to the receptacle terminals.

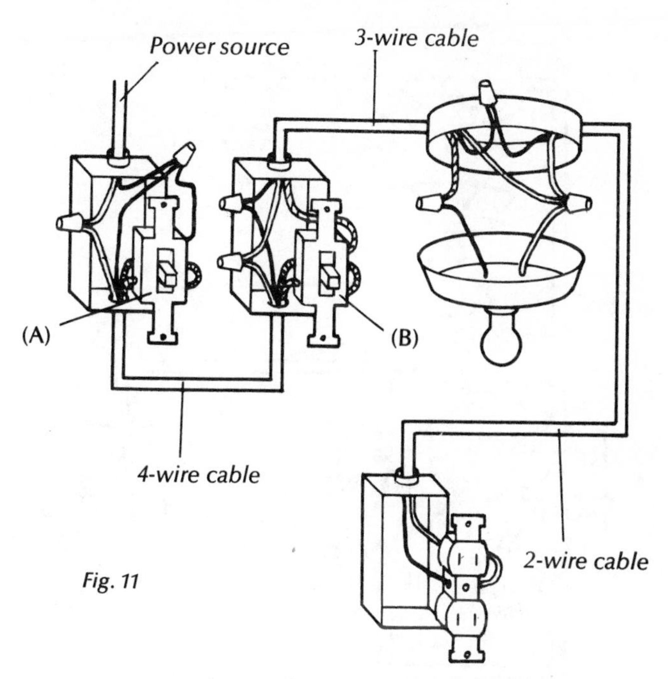


(A) and (B) are three-way switches

## *Controlling one fixture with three-way and four-way switches (Fig. 12)*

You can control a hall light, for example, by placing three-way switches at either end of the hall and intermediate four-way switches in between them. Anytime you add a switch between the two extremities of a run, it must be a double-throw, or four-way, switch. With this kind of arrangement, the power source comes into the fixture itself and can be a two-wire cable. Two-wire cable also connects switch (A) with the fixture, but from then on all of the switches are connected with three-wire cable. At the fixture box, the black wire from the power cable is attached to the black wire from three-way switch (A). The white wires from the fixture and power cable are connected, while the black wire from the fixture is hooked to the white wire from switch (A). The white wire is painted black. At switch (A) the two-wire cable from the fixture enters the box and a three-wire cable leads to the four-way intermediate switch, (B). The white wire from the two-wire cable is connected to the black wire from the three-wire cable. The black wire from the two-wire cable and the red and white wires from the three-way cable all are connected switch (A) terminals. Switch (B) has a three-wire cable leading into it from switch (A) and another one leading out of it to the other three-way switch, (C). The black wires in the two cables are connected, while the

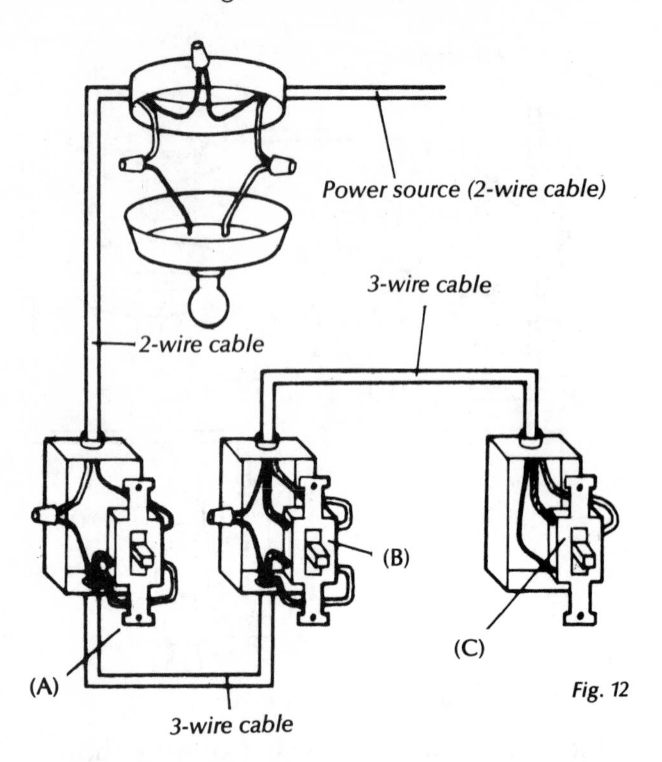


*(A) and (C) are three-way switches; (B) is a four-way switch*

red and white wires from both cables are attached to terminals on switch (B). At switch (C), all three wires in the three-wire cable leading from switch (B) are connected to terminals.

## WHEN THE CIRCUIT HAS BEEN RUN

After the cable has been extended from outlet to outlet and all of the boxes are secured in their holes, the final run of cable from the last box to the service panel can be made. The cable should be secured to the side or edge of the ceiling joists with U-nails and enter the service panel at one of its open positions. Only at that point should you connect the line to the distribution box, or hire an electrician to complete the final hookup. While the electrician is in the house, you might as well have your connections checked. If your work gets by a professional, it will probably pass an official inspection.

### *Connecting branch circuits to a distribution panel*

1. **Pull the main switch to Off. Be careful—all of the cable wires and terminals behind the switch are live even after the switch is in its Off position. Do not touch them.**
2. Take off the front of the distribution box.
3. Remove enough sheathing from the cable so that the wires can reach the terminals in the box.
4. Clamp the cable in the panel box knockout nearest the position you are using.
5. Connect the white wire to a neutral, or grounding, terminal in the panel box.
6. Connect the black wire to the positive terminal in the socket.
7. Attach the bare grounding wire to any unused terminal in the grounding busbar.
8. Turn on the main switch. Test the branch circuit with a hot-line tester (see page 13).

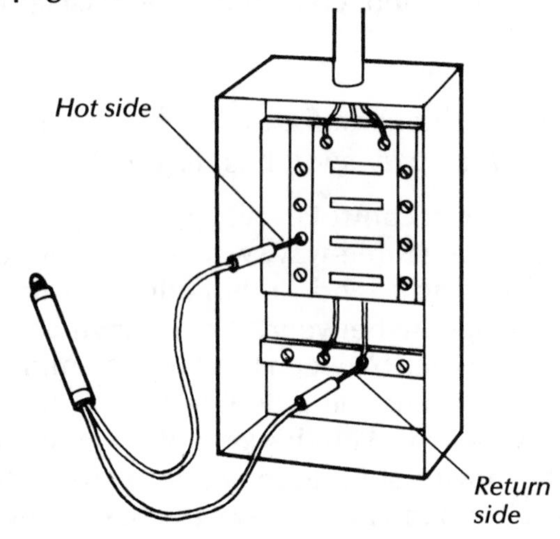


*Check out the new circuit, starting at the distribution box*

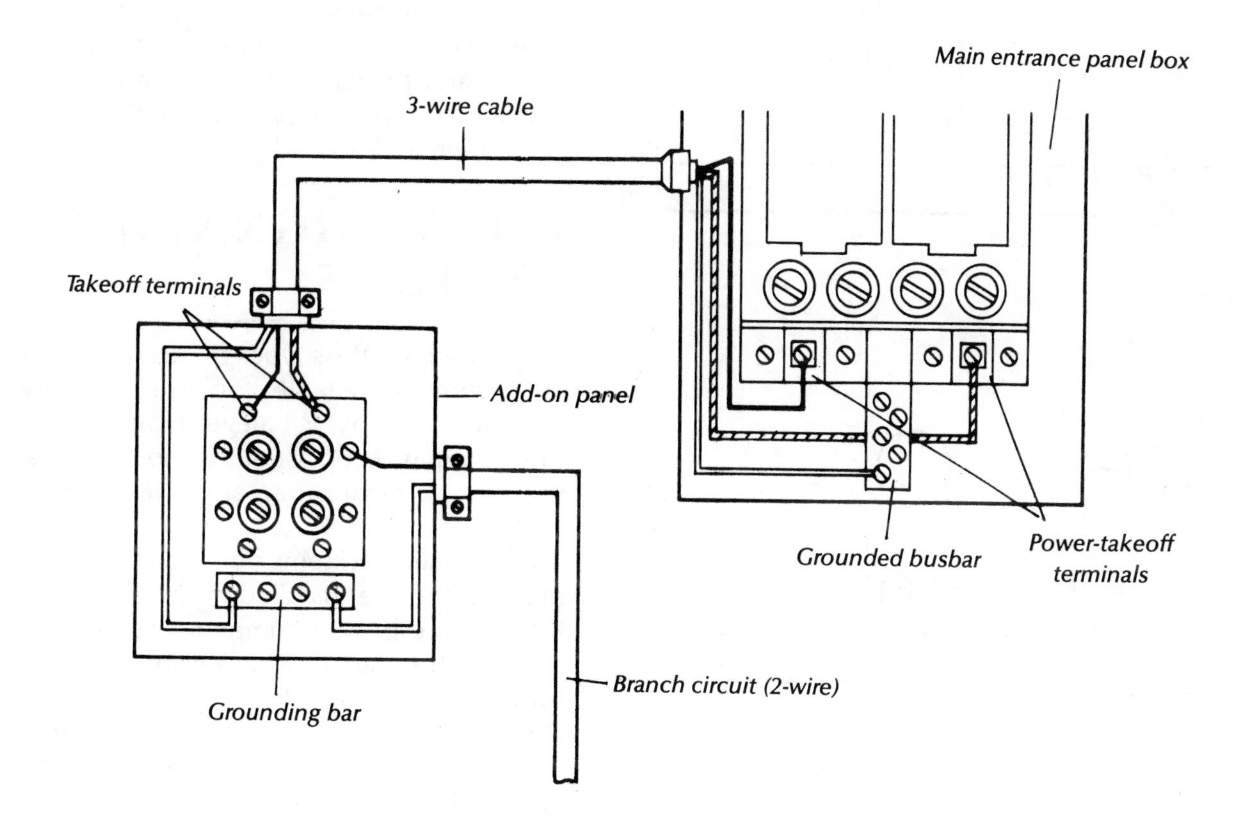


*Wiring of add-on panel box*

9. Replace the front panel.
10. Test each of the outlets and switches in the circuit.

### *Installing an add-on panel*

If there are no more positions left in your distribution panel, you can still connect a four-position add-on panel that will handle four branch circuits.

1. **Shut off the main power disconnect.**
2. Open the main entrance panel box. You will find a power-takeoff terminal located between the two extreme-left fuses and another one between the two extreme-right fuses.
3. Use a #10 three-wire cable to connect the two hot lines to the power-takeoff terminals.
4. The white wire in the three-wire cable is connected to any free terminal on the grounding bar (where all the rest of the white wires entering the panel are connected).
5. At the add-on panel, join the two hot lines to the power-takeoff terminals.

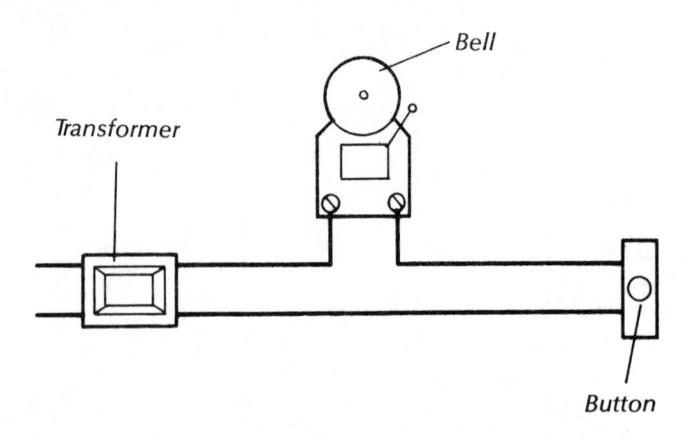


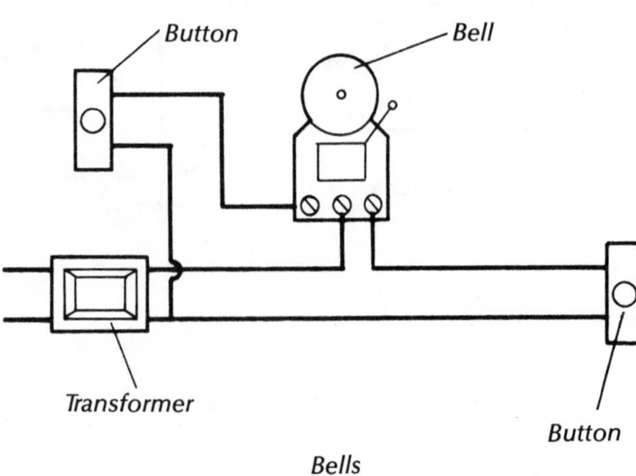


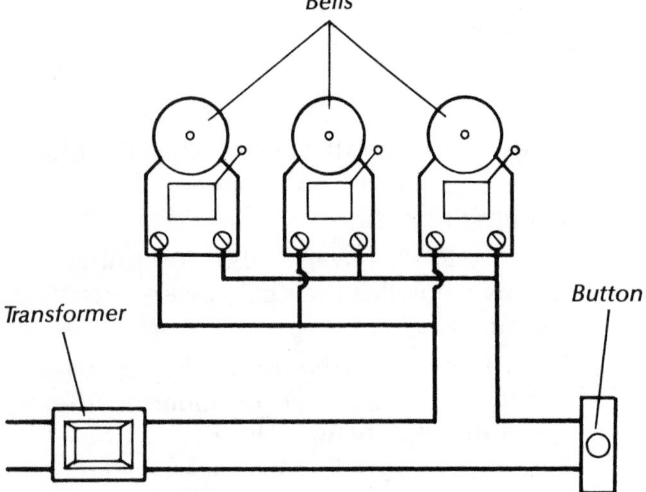


*Three wiring hookups for a signaling system: (top) one bell operated by one button; (center) one bell, two buttons; (bottom) three bells (in different locations), one button*

6. Connect the white wire to the grounding bar.
7. A branch circuit can be brought into each of the four positions in the add-on box and protected by either a fuse or a circuit breaker, depending on the design of the panel box.

# HOUSE SIGNALING SYSTEMS

There are countless types of buzzers, bells, and chimes available for use in house signaling systems. Whatever type you use, the system will consist of a button, a signaling unit, a transformer, and #18 insulated wire.

All signaling systems operate in the same manner and are installed in the same way. Pushing the button completes the contact between the signaling device and the transformer, activating the buzzer, bell, or chime. Bells and buzzers have a clapper that is vibrated by an electromagnet; the electromagnet on a set of chimes attracts a rod, which in turn strikes the musical tubes in the unit.

*Transformers*—These consist of a primary coil (winding) that is connected to the house circuit and receives 120 volts. The current flows through the primary coil to a low-voltage secondary coil, which reduces the voltage to between 6 and 10 volts if the signaling device is a buzzer or bell, to between 12 and 18 volts for chimes. A threaded nipple is attached to one side of the transformer so that the unit can be secured in the knockout of a junction box or the side of a distribution panel. Wires from the primary coil are connected to the house circuit; #18 insulated wires from the secondary coil are used to connect the button and signaling device.

## Signaling System Repairs

The button is responsible for almost every failure in any signaling system—it corrodes, rusts, jams, becomes dirty. Test the button by taking off its wires and touching their ends together. If the signaling device sounds, the button is faulty and should be replaced. You can clean its electrical contacts with fine sandpaper, but you can just as easily install a new button and get years of trouble-free service.

*Signaling unit*—Check to see whether power is entering the unit by touching the probes of a hot-line tester to the unit's terminals. If the tester lights, clean the signaling device. Be sure all mechanical parts are operating. If they do not work easily, replace the unit. If the unit works mechanically and the tester does not light, the transformer is faulty.

*Transformer*—Touch the tester probes to the terminals on the low-voltage side first, then test the 120-volt side. If the light does not go on both times, the transformer must be replaced. If power is getting through the transformer and the system does not work, there is a break in the wiring.

*Wires*—Examine the wires closely for breaks. You can splice broken wires together and wrap the splice with electrician's tape, or replace the wires.

### Replacing a Signaling System

If you want to replace a bell or buzzer with chimes, you must also replace the transformer with one that delivers between 15 and 20 volts.

To change the wires, simply disconnect them and tie them to the ends of the new wires. Pull the old wires out of the walls until the new wires appear. If the wires are in good condition, they do not have to be changed.

#### *Installing a new signaling system*

1. Drill a hole for the button in the door frame. The exact size of the hole depends on the unit you are installing.
2. Drill a hole in the cellar ceiling, through the bottom of the door frame.
3. Fish either single #18 wires or a two-wire cable through the holes. Tack the wires to whatever you can with U-nails until you reach the low-voltage side of the transformer.
4. Connect the wires to the back of the button.

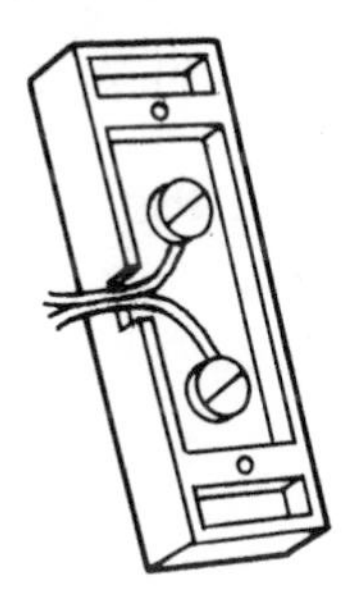

*Connect wires to screw terminals on back of button*

5. Install the button in the door frame.
6. One button wire goes to the signaling device. Connect it to a terminal.
7. A second wire goes from the signaling unit to the transformer.
8. If there is to be a second button at another door, install it as described in steps 1–6. The wire from the second button will be connected to the same terminal on the transformer as the first button. The other wire will be attached to its own terminal on the signaling device.
9. **Remove the fuse or turn off the circuit breaker on the branch circuit feeding the transformer.**
10. Secure the transformer in the knockout of an electrical box or the distribution panel.

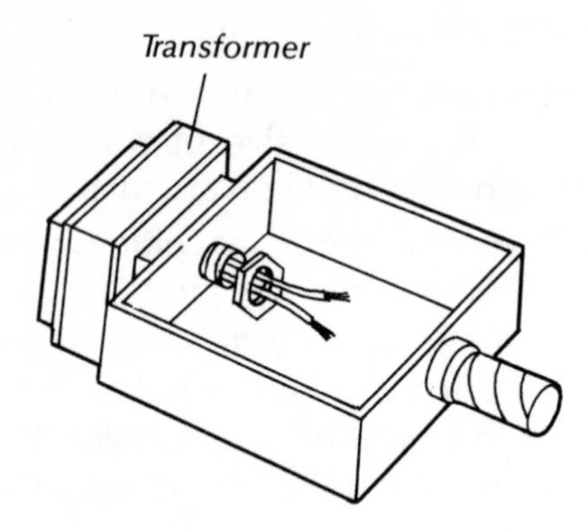


*Attach transformer to box*

11. Connect the 120-volt wires from the transformer to the circuit wires (white to white, black to black).
12. Connect the wire from the button to a terminal on the low-voltage side of the transformer.
13. Connect the wire from the signaling device (step 7) to its own terminal on the transformer.
14. Turn on the power and test the system. If it does not work, check all of the terminal connections and any wire splices you may have made.

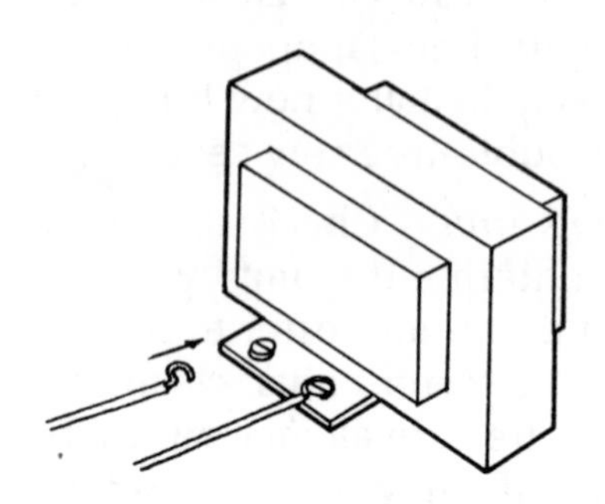

*Attach the button and signaling device wires to the low-voltage side of the transformer*

# CHAPTER 5 • Outdoor Wiring

There are any number of reasons you might want to extend an electrical system beyond the four walls of your home. You might want a system of floodlights placed under the eaves, in and around a garden or a patio, or around a swimming pool. You could need to illuminate a garage, outbuilding, or parking area, install a post light, or add an outlet under the eaves (controlled by an indoor switch) for heating cables strung along the gutters to prevent a build-up of snow on the edges of the roof.

Whatever your outdoor electrical needs, the basic connections and assemblies remain the same as any you might make indoors. The difference comes with the equipment you must work with and the fact that whatever you do must be done with the idea of maximum weatherproofing in the front of your mind.

## Outdoor Electrical Equipment

The cables you are permitted to use outdoors are specially insulated and protected for burial in the ground or to withstand the ravages of nature. Outlet boxes and switches are designed to be weatherproof and must also be caulked after their installation. And you will have to get used to working with thin-walled conduit and its accessories.

### *Weatherproof Boxes*

The outdoor electrical boxes that hold wire connections, switches, and outlets usually have snap-on or screw-on covers over their faces, and because they are designed to withstand the weather, their sides cannot be removed to gang them together. Among the various metal boxes available

*Outdoor boxes must be weatherproof*

are Bakelite wall and ceiling boxes for specific use in damp locations.

You will also find on the market a host of weatherproof light fixtures that are UL-listed for installation outdoors. Included among these are several types of lamps that come with a photoelectric cell that automatically turns an outside lamp on at dusk and off again at dawn.

## Outdoor Wire

Use #14 or heavier wire when you are making exterior connections, but the best approach is to consult your local building code. It will most likely permit the use of service entrance cable, which has a moisture-resistant, flameproof covering around stranded wires that are #6 gauge or larger. Another widely accepted wire is underground feeder cable, which can not only be used outdoors but also can be buried underground without being run through a conduit. A third alternative is lead-encased underground cable, which is normally used to bring power to outbuildings from a main power source, although many local codes now forbid its use.

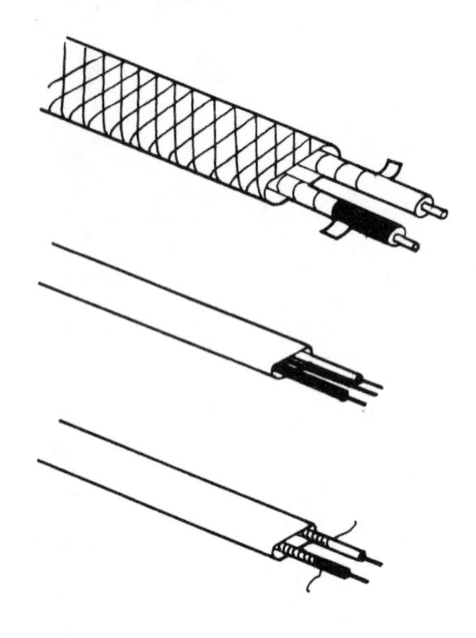

*Protection is needed for outdoor wiring*

## Conduit

Thin-walled metal or plastic conduit is pipe. It is sold in 10′ lengths and various diameters, and has its own special array of connectors, adapters, fiber rings, and clamps designed to connect lengths of conduit in a series of waterproof seals. When you have assembled the conduit to go from just inside your house to wherever you want the electrical wiring to run, you must then pull the cable through it. One of the advantages of metal conduit is that it acts as its own grounding conductor. Although conduit presents an added expense, in return you do get a certain amount of assurance that probably nothing will ever damage the wiring inside it.

Because it is thin-walled, conduit can be

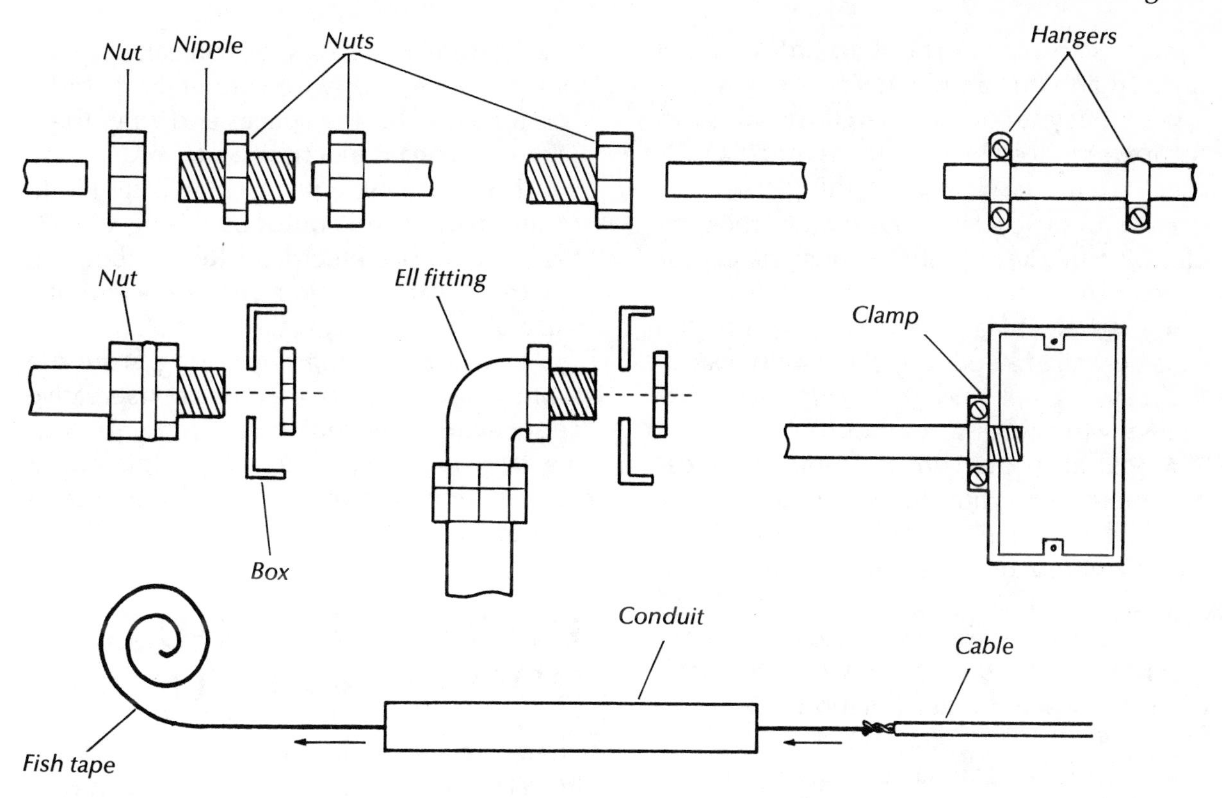


Conduit and some of its accessories

made to bend around corners. You may not, however, have more than 360° of bends in any conduit between two electrical boxes.

The tool that bends conduit is called a hickey. The hickey is a handle attached to a curved track with an adjustable clamp at one end of it. Hook the bender over the end of the conduit, then pull the handle back in the direction you want the bend to go. It helps if you stand on the conduit, and stop occasionally to check the angle of the bend.

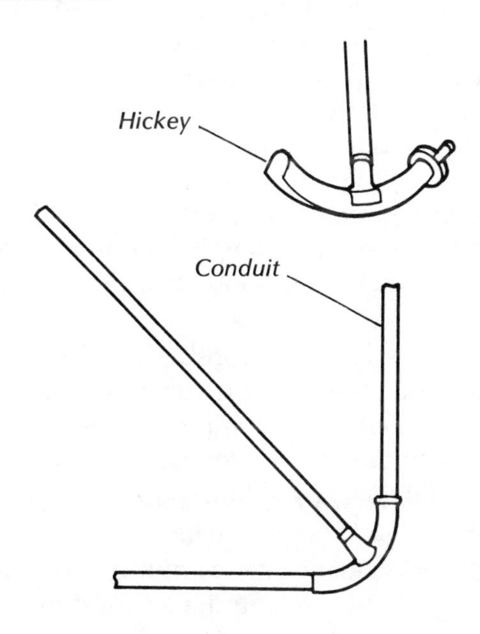


Bending conduit with a hickey

## SOME RULES FOR OUTDOOR WIRING

Any exterior wiring is vulnerable to the weather, particularly any form of dampness, which can short out an entire circuit faster than you can blink. The rules to follow so that your outdoor circuitry will serve for years without trouble are as much a matter of common sense as anything else.

- Always use a UL-listed switch, receptacle, or junction box that is designated for outdoor use.
- If an electrical box is mounted to the outside wall of a building, the joint between the cover plate and the wall should be sealed with caulking compound.
- Seal all of the unused holes in face-mounted boxes with threaded plugs.
- The connection between an outside-wall-mounted box and the inside of the house should be made with a length of conduit, whether the rest of the wiring uses it or not. The conduit is screwed to the back of the box and ends at a junction box inside the house.
- Any tool or appliance plugged into an outside outlet should be double-insulated or fully grounded. Any light bulb used outdoors must be weatherproof; standard incandescent bulbs can shatter when hit by rain or snow.
- Do not do any outdoor wiring without consulting your local building code for designated equipment and proper installation requirements. For example, you may be required to install a ground-fault circuit interrupter.
- Use only waterproof cable with outdoor wiring, even if it is run through conduit. The local code will tell you exactly which cables you are allowed to use.
- Anytime your cables must run along the outside of a wall, or wherever there are sharp rocks or any chance of the wire's becoming frayed or severed, use conduit.
- Never run your cables through a doorway or window.
- In some communities you can use wire nuts to make your wire connections, but it is safer to solder the splices and wrap them with waterproof electrician's tape.
- Even when you just change a light bulb in any part of the outdoor wiring circuit, first shut off the electricity. Remember that electricity can give you a shock—and any trace of water could make it lethal.
- If you are doing any wiring around a swimming pool, be overly conservative about everything you do. It is better to over-insulate, double-protect, and triple-check every connection than to risk water getting at your wires.

## INSTALLING OUTDOOR CIRCUITS

There are two kinds of outdoor electrical installations: those fixed directly to an outside wall of your house and those extending a circuit to some sort of fixture away from the house. In both cases you are taking power from the distribution box inside your house and bringing it outside. If at all possible when running an entire branch circuit outside the house, connect it to its own GFCI installed in the power distribution panel or an add-on panel box. Exterior-wall outlets or fixtures can be connected to an existing indoor branch circuit in a junction box.

### *Tools*

When you are working with an outside circuit you will need the same pliers, screwdrivers, and strippers used with any wiring project. You will also need an electric drill and a utility saw in order to cut through the exterior siding of the house,

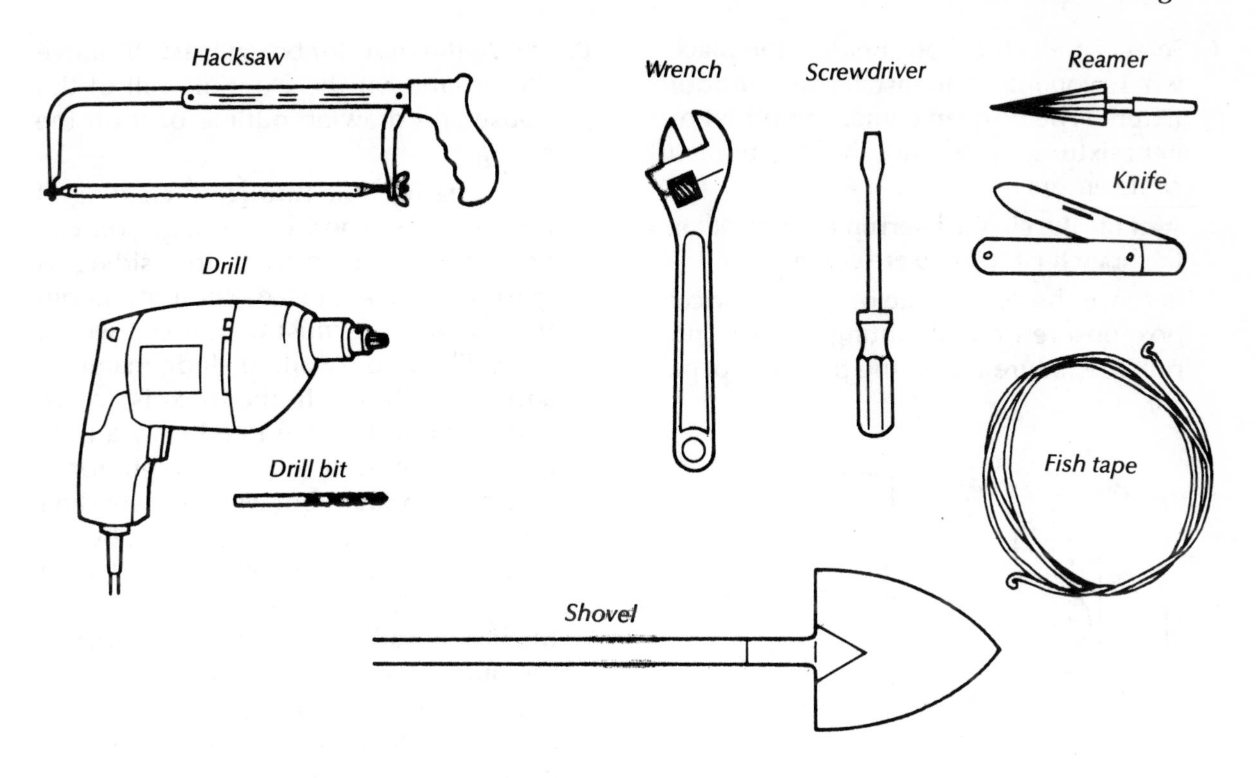


*Some of the tools used for outdoor wiring*

and fish tape for pulling wires through conduits as well as through interior walls.

## *Materials*

Aside from the normal junction boxes for use indoors, also purchase weatherproof outdoor boxes with protective cover plates and gaskets and a good weatherproof caulking. Use whatever outdoor cable is permitted by your local building code and whatever conduit and fittings you need to complete the job at hand.

When you are calculating the length of cable and conduit, add 10′–20′ to your computations to allow for burial underground. Your estimates should be taken from the farthest fixture outside the house to the power source indoors.

## *Installing fixtures on an outside wall*

When you plan the installation of outside wall outlets, try to place them so that you will not need to use extension cords. It is safer to install a few extra outlets that let you avoid using extension cords, which can fray and then become damp and cause a dangerous shock.

Your wiring should run from an inside source of power as directly as possible to an opening in the outside wall, which means you may have to fish cable from an outlet or fixture through walls, over ceilings, under floors, or along baseboards.

1. Select an electrical box nearest the place where you intend to install your outdoor fixture. The box can contain an outlet or light fixture, even a switch if the branch cable enters its box.
2. **Turn off the circuit interrupter controlling the branch circuit you are working on.**
3. Remove the cover plate on the electrical box, unscrew the mounting screws, and pull the fixture out of the box (see page 26).

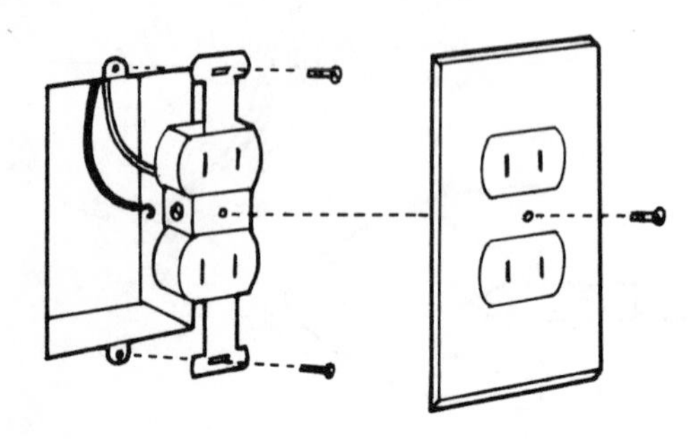

*Remove the indoor fixture from its box*

4. If necessary, remove the electrical box from inside the wall in order to run cable from it to your outdoor fixture position.
5. Remove one of the knockouts in the side of the box that faces in the direction you want the cable to run.
6. Install a cable clamp in the knockout and insert one end of the cable. Strip the cable wires and prepare the bared ends for connection in the box.
7. Fish the cable to the inside location of the outdoor box.

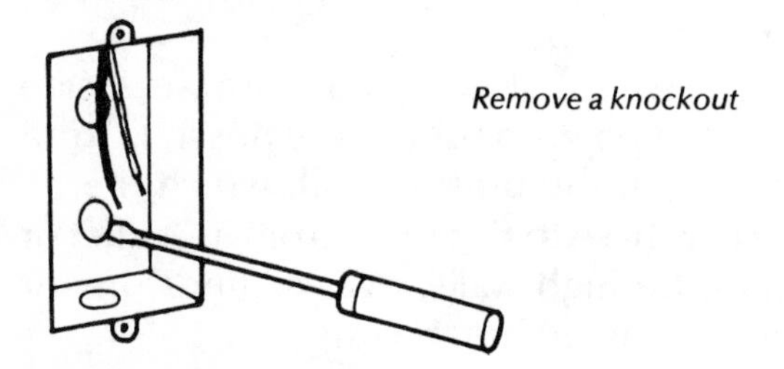

*Remove a knockout*

8. Locate the outdoor box at least 18″ above the ground on the exterior wall of the house and draw an outline of it on the siding.
9. Drill a starter hole through the siding. If the siding is wood or asphalt, you can use an electric drill. If the siding is masonry (brick, stucco, cement, stone) the box outline must be cut out with a star drill, masonry bit, or sledgehammer and cold chisel. If the box is to be surface-mounted you need only a hole through the wall large enough for a length of conduit, but when embedding the entire box, its rectangle must be cut through the wall. The outside wall most likely has a hollow space between it and the inside wall. If you are face-mounting the outside box, when you have cut the rectangle in the outside wall, position the box in the hole and drill through its back-side knockout to the interior wall directly behind it. If you are surface-mounting the outside box and drilling only the center hole, continue it through both the outside and inside walls.

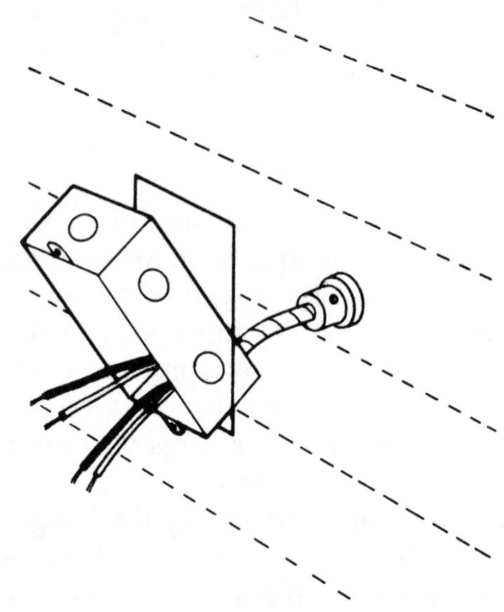

*Cut a box-size rectangle in the outside wall*

10. You have the option in many locales of simply running your power cable through the outside wall to the electrical box, in which case the cable from your inside electrical box must be rated for use in damp locations. A safer method of making your connection, which is required in many communities, is to connect the outdoor box to a box positioned directly behind it on the interior wall and connected to it with a length of conduit. The interior box can be a covered junction box, might contain outlets, or, better still, might have a switch controlling the outside fixture.

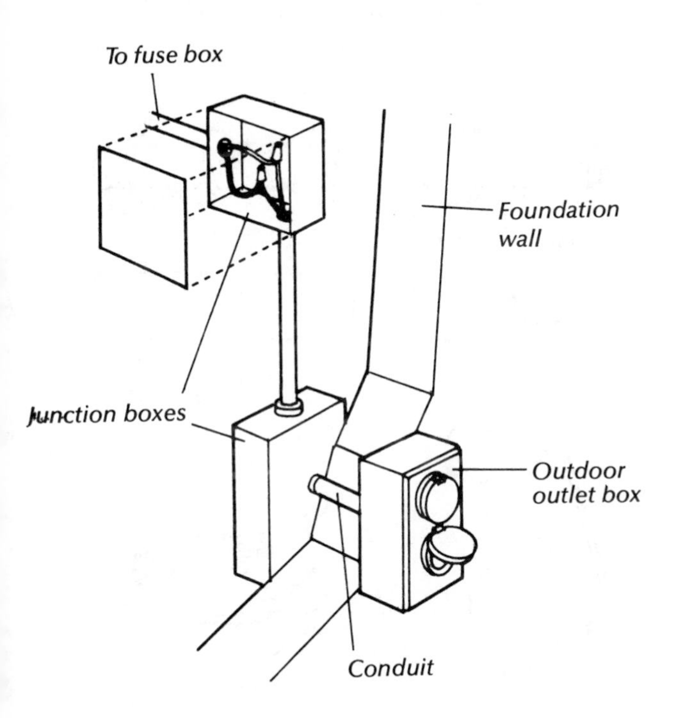


Connect the outdoor box to an indoor box

To assemble the electrical box arrangement, first affix a length of conduit to the back knockout in the outdoor box.

11. Affix the outdoor box to the exterior wall so that the conduit extends through the wall to the inside of the house.
12. Cut the necessary hole in your inside wall for a junction box and install it in the wall. The conduit should be screwed into the knockout in the back of the box.
13. Lead the power cable from the takeoff box into the junction box and secure it with a cable clamp. Strip the cable wires and prepare them for connection.
14. Insert a length of weatherproof cable through the conduit and strip the wires at both boxes, then prepare them for connection.
15. Install the outlet, switch, or fixture in the outdoor box, connecting the black wire from the conduit to a brass terminal and the white wire to a silver terminal.

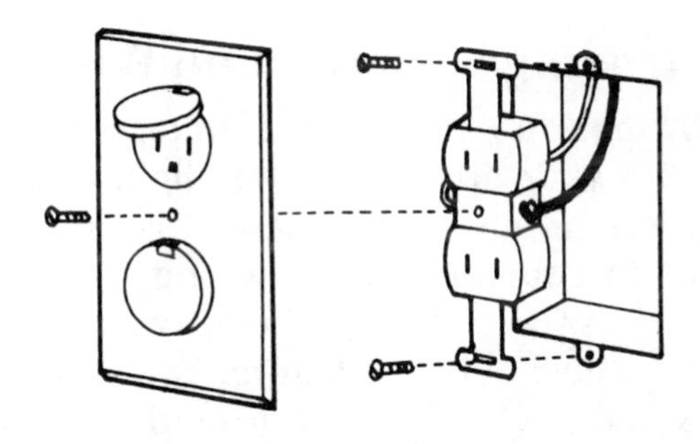

Wire the outlet

16. Connect the cable grounding wire to the electrical box.
17. Secure the fixture to the box (see page 27).
18. If the box contains a switch or outlet, secure the rubber gasket over it and attach the weatherproof cover plate.

19. Apply caulking around the sides of the electrical box.
20. At the inside junction box, either connect the wires from the conduit and the power source to their proper terminals on a switch or receptacle, or splice them together with wire nuts.
21. Install the switch or outlet. Connect the cable and fixture-grounding wires to the electrical box.
22. Secure the fixture in the box.
23. Install the cover plate.
24. Connect the takeoff cable wires to their proper terminals or wires in the takeoff box.
25. Connect the takeoff cable's bare grounding wire to the takeoff box.
26. Install the fixture back in the box and attach its faceplate.
27. Turn on the circuit interrupter and test all of the fixtures that you have connected.

## Extending Underground Wiring

When you want to wire a lamp post at the end of your driveway or spotlight a patio, garden, or any other area that is not adjacent to your home, the wiring must run from its source of power in the basement to the farthest outlet or fixture. You can bury the cable 6″–18″ belowground and not bother to run it through conduit, but conduit provides additional protection against dampness and damage from people and machines.

### *Tools*

With the normal complement of electrical tools (pliers, screwdrivers, fish tape, strippers), plan to have available a hacksaw, pipe reamer, adjustable open-end wrench, hickey, and shovel, all of which are needed for working with conduit.

### *Materials*

You will need a full complement of electrical boxes, fixtures, connectors, couplings, ell fittings, and clamps for whatever conduit you are using. Be certain that all of the fittings you buy are the proper size to fit the conduit.

Unless you are adding only two or three fixtures or outlets, it is best to give your outdoor circuit its own protective device in the distribution panel, preferably a GFCI. If you are connecting the outdoor circuitry to

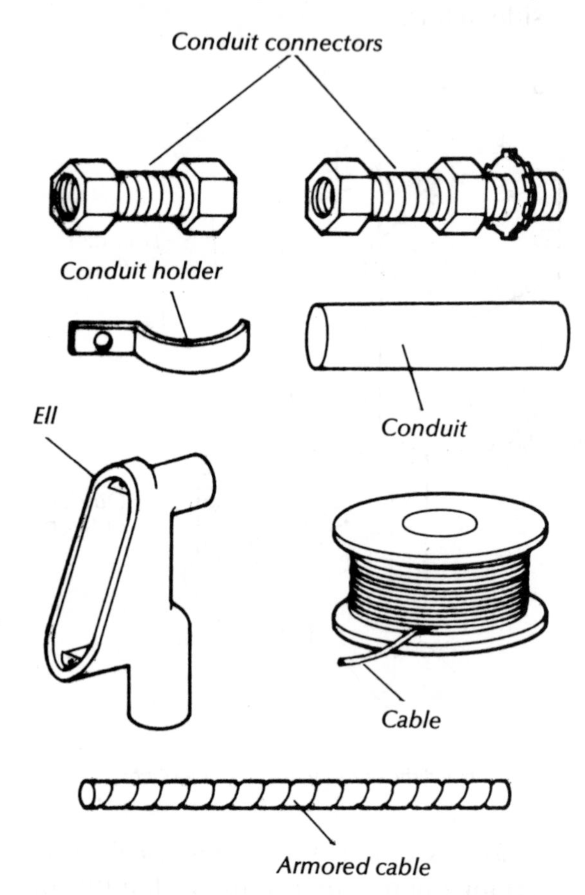


*Materials needed to install an outdoor branch circuit*

an existing branch circuit inside the house, the connection can be made following the procedure on pages 80–82, except that the hole opened through the outside wall need only be large enough to accept a conduit ell fitting.

1. Beginning at the distribution panel, attach weatherproof cable to the basement ceiling from the panel box to the point where it will exit the basement.
2. Drill a hole through the exterior wall to accept an ell conduit fitting.
3. Install the ell fitting in its hole and push the end of the interior cable through the back of the ell. Remove the cover and gasket screwed to the back of the ell.
4. Cut the cable and strip its wires. The wire connection between the indoor power cable and the outdoor cable will be made inside the ell.
5. Beginning below the ell fitting, dig a trench directly to the location of the outdoor fixture. The trench can be as shallow as 6″ belowground, but to protect it from gardening tools, erosion, and frost, the recommended depth is 18″. As you dig the trench, carefully cut the sod into blocks and lay them on pieces of burlap so they can be replaced after the conduit is laid and the trench filled in.

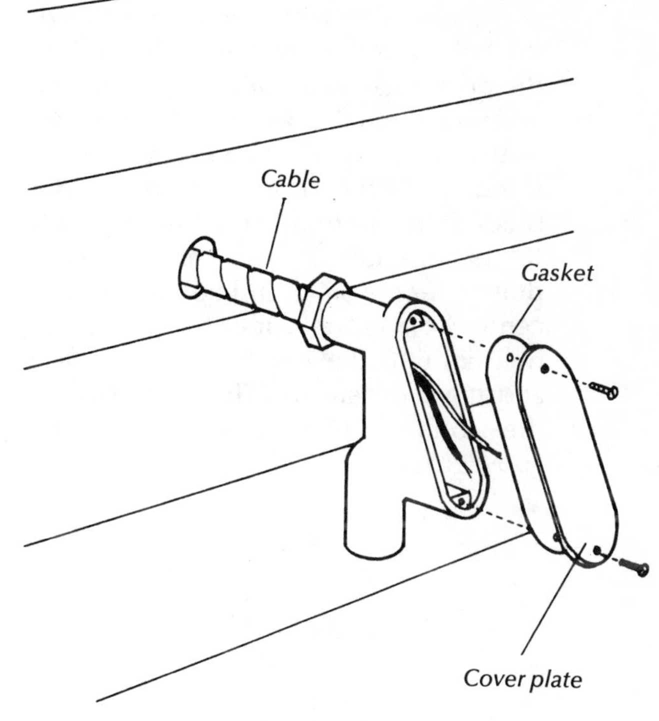


*Installing an ell fitting*

6. At the end of the trench, the fixture must be secured in the ground. If you are installing a post lamp, dig a hole 3′ deep

*Conduit run underground to an outlet away from the house*

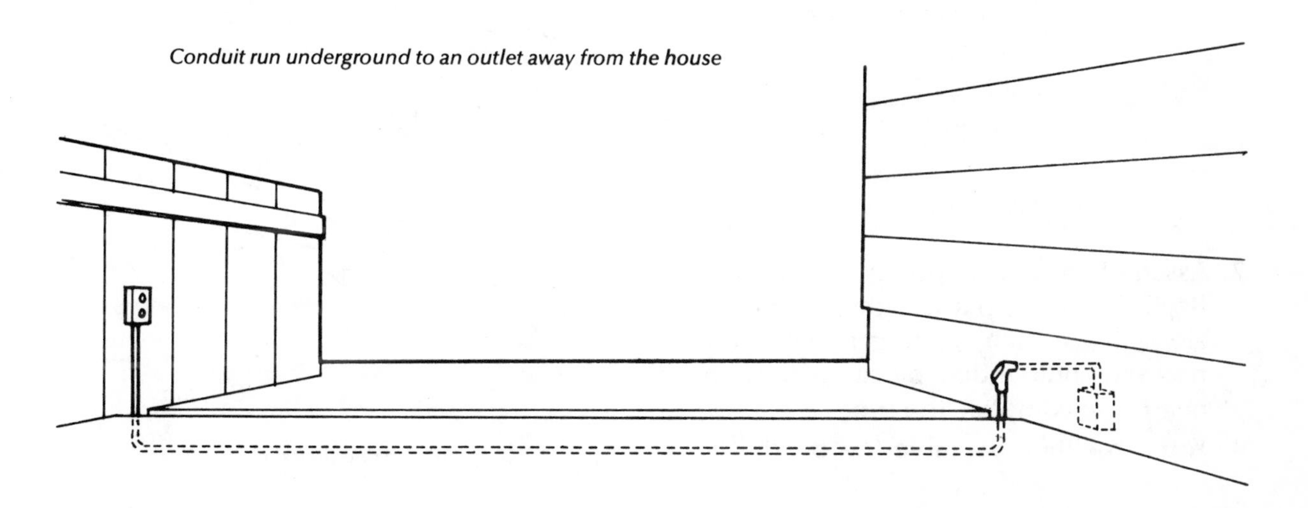

and about 18″ square. Line the bottom of the hole with 3″–4″ of gravel, then install the post, bracing it upright with stones, and fill around the debris with concrete.

An easy way of stabilizing a free-standing electrical box is to bury a cinder block at the bottom of the trench (if it is 18″ or more deep). Then run your conduit under the block and up through the center hole in the cinder block. Fill the hole around the conduit with cement and attach the electrical box to the top of the conduit, placing it at least 18″ aboveground.

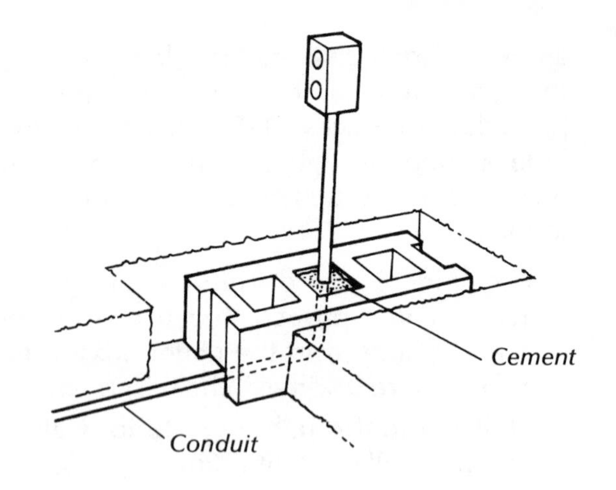


*A cinder block is a stable base*

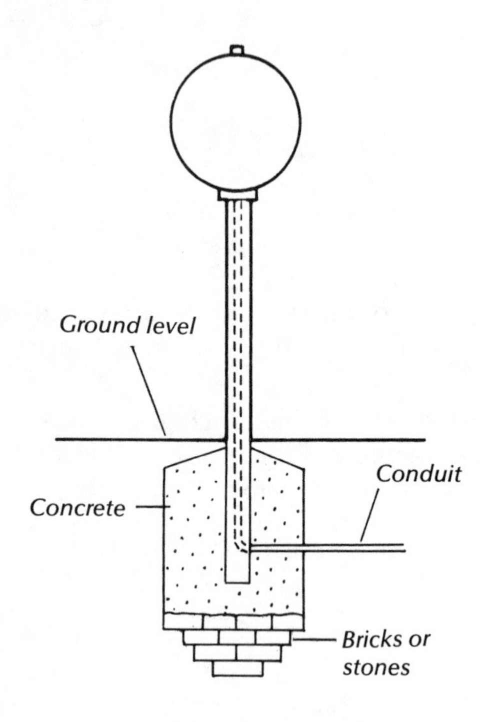


*Installation of post lamp*

7. Assemble the conduit in the trench. Begin by laying out lengths of conduit along the trench, cutting it with your hacksaw and bending pieces as they are needed to complete the run.
8. You can either assemble the conduit in its entirety and then fish cable through it or string lengths of conduit on the cable and assemble it section by section. Begin at the ell fitting by removing the locknut on a connector (it is not needed for making connections to ell fittings).

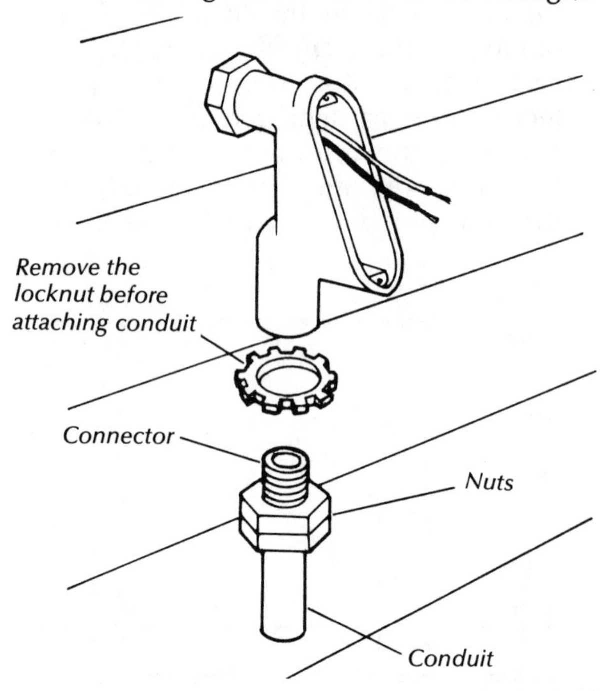


*Attach conduit to the ell*

Screw the connector to the ell fitting and tighten its nuts.

9. The first section of conduit should run down the outside of the house wall to the bottom of the trench and make a right angle along the bottom of the trench. When it is connected to the ell fitting, insert the cable and bring it up into the ell fitting. Strip the wires and prepare the bared ends for splicing with the inside cable entering the fitting.
10. The successive sections of conduit are assembled with couplings between lengths, which are secured by tightening the nuts on the coupling. If you are carrying the cable along with the conduit, slide the coupling around the cable, then fish the cable through the new section of conduit before making your connections.
11. Continue assembling the conduit and fishing cable through it until you have completed the run from your house to the fittings. **You may not make any wire connections inside the conduit.** All wire connections must be made in junction boxes in the middle of your conduit run wherever it is necessary to make wire splices.
12. Make all wire connections to the fixture and inside the ell fitting. Check your local code for an explanation of how you may make wire connections underground in your area. You may be permitted to just use wire nuts, or you may have to solder your splices and wrap the connections in waterproof tape.

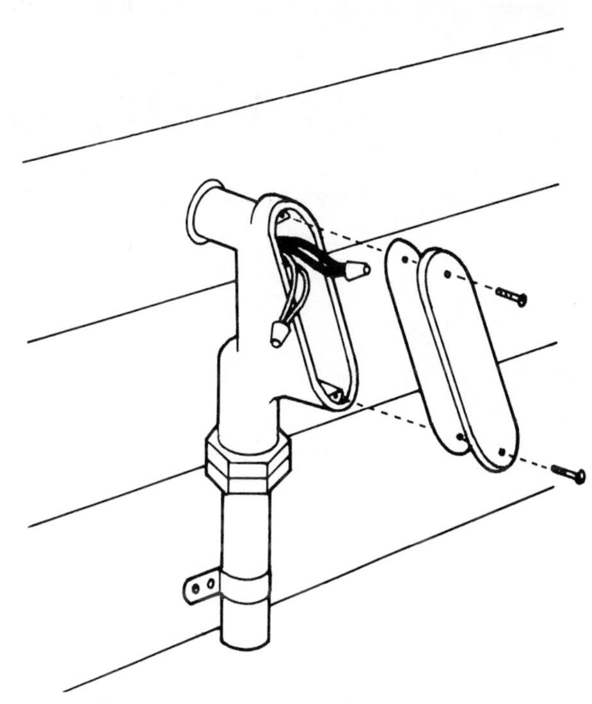

*Connect the wires and cover the fitting*

13. Cover the ell fitting with its gasket and plate.
14. Apply caulking around all fittings where they meet an outside wall of the house.
15. Connect the inside cable to the distribution panel. (See pages 56–57 for instructions concerning connections of a fuse, circuit breaker, or GFCI.)
16. Test the circuit to be sure the fixture is working properly.
17. Fill in the conduit trench and replace the sod squares. Water the sod regularly for two or three days, or until the grass no longer shows any scars from having been dislodged.

# CHAPTER 6 • Appliance Repairs

Usually, when something causes the fuse or circuit breaker on a branch circuit to shut off the current, the problem is somewhere in one of the light fixtures or an appliance plugged into the circuit. If you open an appliance, invariably it appears to be a confusing system of wires leading to and from dozens of components. Obviously, the more components, the more chances for something to go wrong. But all of those components are also small in nature and may be specially designed for only the appliance where you find them. And therein lies the essence of appliance repair: replacement of parts. The majority of repairs made to both large and small electrical appliances consist of locating a faulty part and then replacing it, because it is usually easier to change parts than attempt to fix them.

## Large Appliances

The large appliances in your home include the hot water heater, refrigerator, oven or range, dishwasher, clothes washer, and clothes dryer. Every manufacturer offers the consumer a variety of model designs and constructions. If you counted up all of the pieces and parts in all of the different models of the many large appliances on the market today, you would arrive at a number in the tens of thousands. Nevertheless, all of the large appliances also have some common elements. They all have an outer shell. If they are heat- or cold-producing machines, they will have at least one thermostat. Most have a timer of some sort, and an electric motor. If they use water in any way, they have a pump. There are always switches, cord sets, and wiring harnesses.

All of these basic components are precisely constructed, and each is made up of dozens of parts, any of which can fail at any time. In almost every case it is less expensive to simply replace the entire unit than to repair it. For example, it costs about $90 to call in a repairman to check your washing machine, discover that the timer is bad, and replace it with a rebuilt timer supported by a one-year guarantee. The rebuilt timer costs about $30, and the repairman will take your old timer away as a trade-in. But a brand-new, unused timer costs $90 all by itself and is not guaranteed for any longer than the rebuilt one, and to send your broken timer out for reconstruction would cost you about $40 or $50.

Since most appliance repairmen perform the majority of their chores by replacing components, it has dawned on a large number of consumers that they can save considerable expense by replacing components all by themselves. At which point the cost of replacing your washing machine timer becomes $30 for the part (plus your old timer), plus no more than one hour's working time. Timers, thermostats, motors, and pumps can all be easily replaced in almost any large appliance by practically anyone who can read and is not color-blind.

The parts for large appliances can be procured by reading the Yellow Pages of your telephone directory and are probably no more than a few minutes' travel time away from your home; there are appliance parts stores in nearly every moderately large community. When you enter one of these stores in search of a particular part, be prepared to tell the salesperson the manufacturer's name, the appliance's model number, and the name of the part and its identifying number, which is either stamped on the part itself or is listed in a parts schematic in the owner's manual.

Parts stores have sets of microfilms containing the schematics and parts lists for every major appliance sold in America. Given the above information, the store can locate the make and model of the appliance you own, identify the part of a comparable replacement unit, and retrieve that part from its inventory. You can also probably get some tips and hints about installing the part from the salesperson; the people who work in appliance parts stores tend to be knowledgeable repair experts in their own right.

If the parts store does not have the part you need in stock, you may have to wait a week or so while they order it from the manufacturer, but you would have to wait anyway if you hired a repairman, unless he represented the specific company that made your machine.

Installation of a large appliance part is strictly a matter of reading the service manual. The steps you must follow are usually given clearly, and there are probably drawings and photographs to help you understand what you are doing. The service manuals for many large appliances can be purchased at replacement-parts stores. They can always be bought from the manufacturers. Quite often when you buy a large appliance, you can get a service manual from the outlet that sells you the machine. No matter how you acquire it, if you plan to do any repair work on a major appliance, be sure you own its service manual. You may be able to get it free of charge. More likely you will have to pay up to $10 or $15—but you cannot make any repairs in a major appliance without one.

## Small Appliances

Small appliances include every electrical machine in your home, from blow dryers to toasters to mixers. They are divided into two categories, heat-producing and motor-driven, although some of them, such as a blow dryer, may be both. Heat-producing appliances include toasters, electric blankets and heating pads, electric fry pans and similar cookers. Motor-driven appliances include mixers, blenders, food processors, shavers, can opener/knife sharpeners, vacuum cleaners, and power tools.

Most repairs to small appliances are also a matter of replacing parts. There are hundreds of small appliances in existence, and dozens of versions of each type. The number of parts that a parts store would have to stock is just too many for an average sale that would probably be under a dollar, so there are few small-appliance-replacement-parts outlets other than the manufacturers themselves.

Thus, if you need to replace a part inside one of your small appliances, you have to contact the manufacturer and tell the company which part you need, then wait two or three weeks to receive it in the mail (often at no charge).

The service manual for a small appliance may consist of an exploded drawing of the unit with its parts listed or numbered. You may or may not find that drawing as part of the owner's manual that came with the unit. If you do not have it, you must write the manufacturer for it, or plunge into the unit on your own and pray that you can make detailed enough notes to get the machine back together again.

# MAINTAINING APPLIANCES

Every appliance must be maintained in peak operating condition if it is to function efficiently. It should be kept clean and in many cases periodically lubricated, all of which can be done without ever dismantling the machine.

Anytime an appliance fails to operate properly, there are four immediate steps to take:

1. Check the plug. It should be seated firmly in its receptacle.
2. Test the On-Off switch.
3. Check the fuse or circuit breaker to be sure it has not blown or tripped. If the circuit interrupter is off, there may be a short circuit in the appliance. Restore power to the circuit and turn on the appliance. If the circuit interrupter goes off again, suspect a short circuit.
4. If the appliance uses water, be sure the water can enter the unit, which means check the faucets.

## Disassembling

If you have double-checked all of the exterior possibilities and the appliance still does not work, you can begin to disassemble the unit. But never take apart any more than you have to, and do that only a little bit at a time. Unplug it and proceed in this manner:

1. Start with the obvious—examine all of the dials, knobs, and buttons. Remove

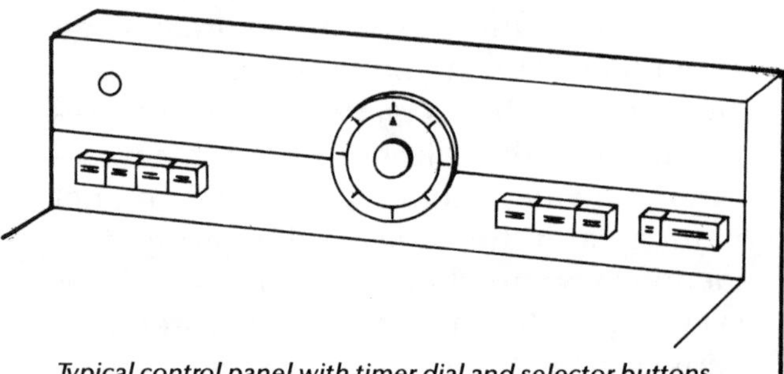

*Typical control panel with timer dial and selector buttons*

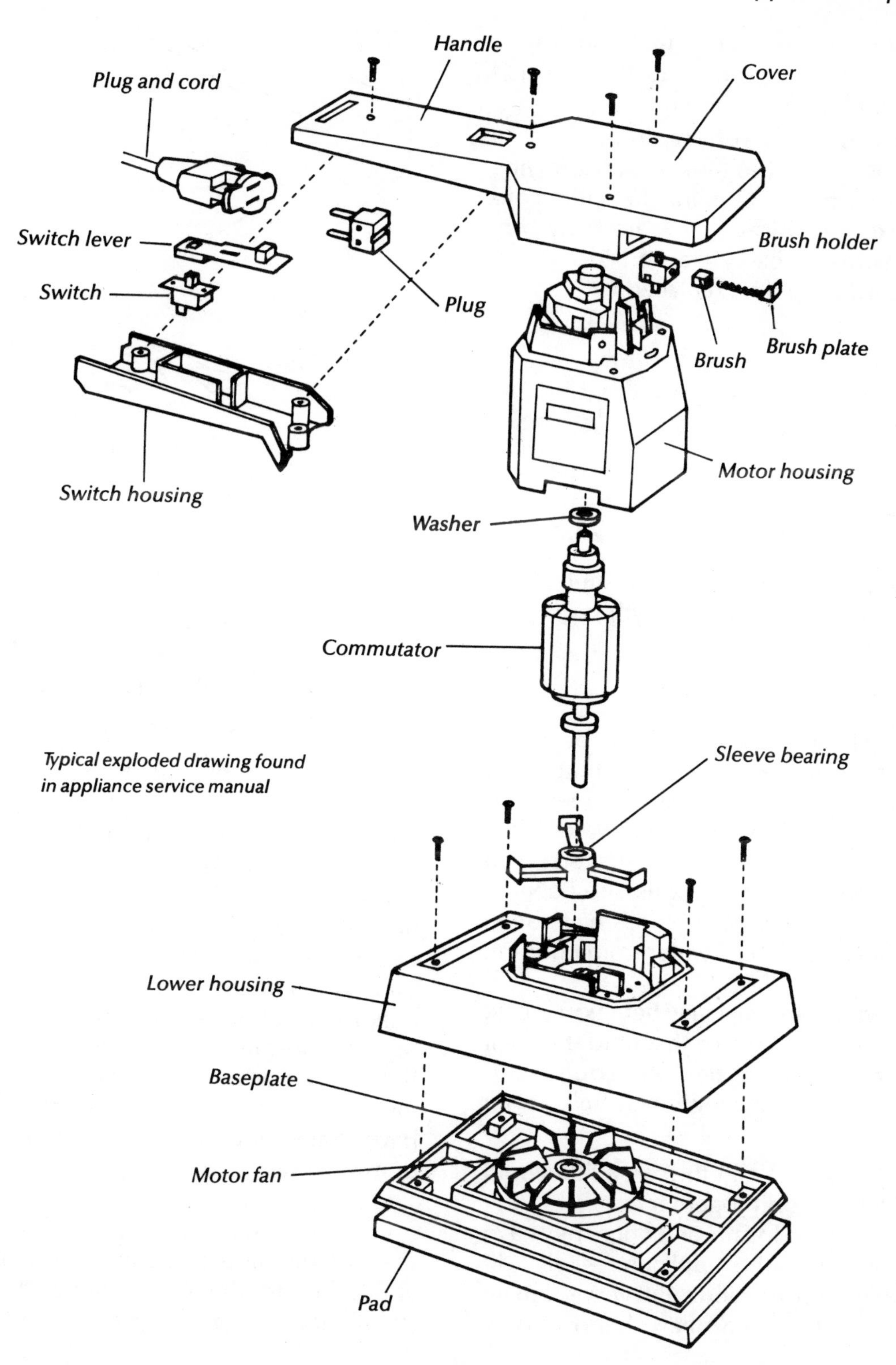


*Typical exploded drawing found in appliance service manual*

any dirt, grease, or grime from them. Work each component to be sure it is functioning mechanically.

2. Check the cord very carefully. Look at the base of the plug for cracks or fraying of the cord wire. Examine the entire cord for exposed wires. If you find any broken wires or badly cracked insulation, the cord should be replaced.

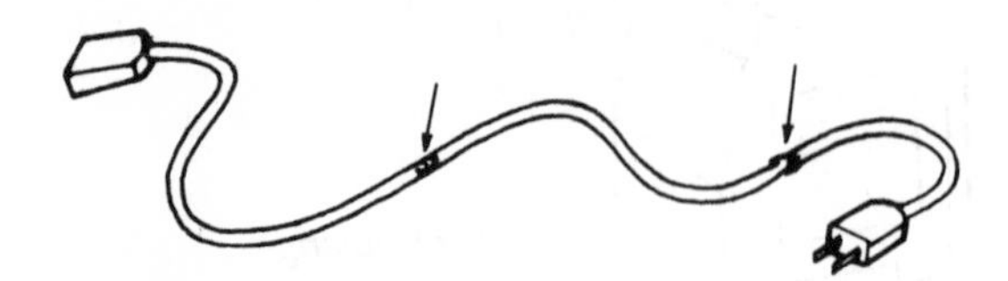

*Look for fraying and cracks in the cord*

3. Look at every surface in the shell of the unit for cracks, holes, anything unwarranted.

## Removing the Outer Shell

Getting inside an appliance can be a frustrating experience. Be assured that every appliance, with the exception of a few hermetically sealed oral irrigators (Water Piks), electric toothbrushes, and shavers, can be opened. The exceptions are labeled "Sealed," and their outer shells have been glued together to keep water from entering the unit.

The owner's manual may help you locate the screws, clips, or tabs that hold the shell together around the electrical equipment. But don't bet on getting much help, even from the service manual.

*Screws*—Sometimes the screws that hold the halves of a unit together are out in the open and easy to locate, and once they have been removed, the appliance shell will come apart in your hands. But very often all or some of the screws have been artfully

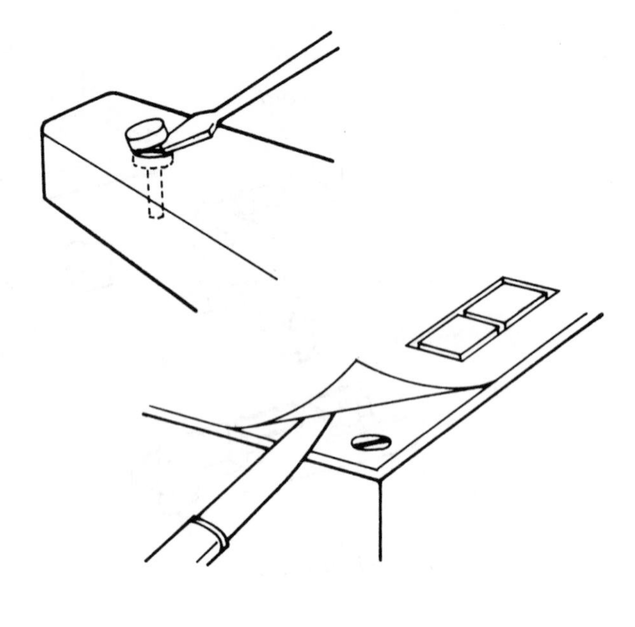

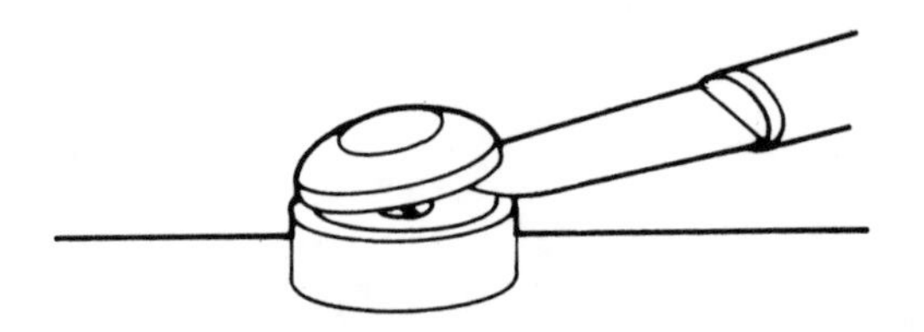

*Screws may be hidden under a plastic cap, decorative label, button, or knob*

hidden in the feet of the unit or beneath round plastic caps that stand out on the surface of the unit as little circles that must be pried free with a knife. Or there may be a decorative panel pasted over the face of the unit, which must be peeled back at its corners to reveal the screw heads. There may be a setscrew in the center of a knob that is hidden by a small cap that has to be pried loose. No matter how they are hidden, you have to find and undo them all before the unit will disassemble.

*Clips*—Spring steel clips are often used to hold the side and top panels of large appliances to the unit's frame. They are usually found at the corners of the panels

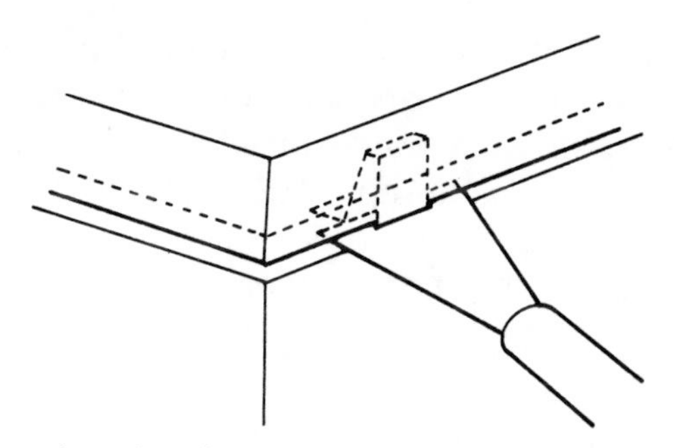

*Often the sides of major appliances are held together by spring clips near the corners*

and can be released if you shove a screwdriver blade into the joint between the panels and push hard. If you can find no other means of assembly, probe for the clips, beginning at the corners under the top of the unit.

*Knobs and dials*—Knobs and dials are attached to some kind of shaft and must be removed before the shell behind them can be taken off the machine. A surprising number of knobs just pull off their shafts. If pulling gets you nowhere, look at the base of the knob or dial for a setscrew to loosen, or a small octagonal hole that will indicate

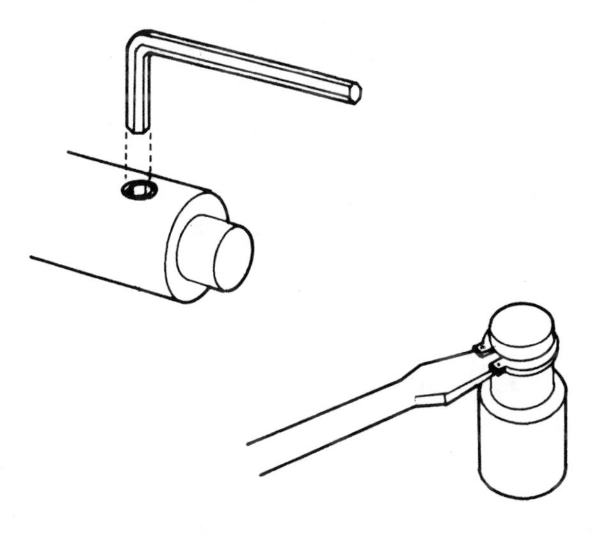

*Disassembly may require the use of an Allen wrench or a screwdriver*

an Allen screw. To loosen the Allen screw you need a set of Allen wrenches, one of which will fit into the hole and can be rotated counterclockwise to loosen the screw. Knobs and dials can also have a spring clip that you release by pushing against it with a screwdriver blade while at the same time pulling on the knob. Or, if they are threaded to their shafts, rotate them in the direction opposite to the way they are turned in use.

## When the Shell Has Been Removed

Once you have gotten the shell of the appliance apart, follow this procedure:

1. Look. Touch nothing—just look. If you have a Polaroid camera, take some close-up pictures of the entire unit; otherwise, make notes and sketches of where wires are attached and how various parts are assembled. If you look very closely, you may discover that the colored wires are connected to terminals on parts that are labeled with embossed numbers or letters.
2. When you are certain that you have recorded enough information to get the appliance back together again, remove one component at a time. Examine each piece for wear or breakage. If it can be tested with a multimeter (see page 94), test it. If it is dirty, particularly its electrical contacts, clean it. Contacts can be cleaned with silicone electrical contact spray. Manually work the part to be sure it does not bind. If the part does not appear to be defective, replace it in the appliance and continue on to the next part.

Be careful never to use excessive force. All appliance components are closely machined, and if they become stuck in any way, you can assume the fault is not in the

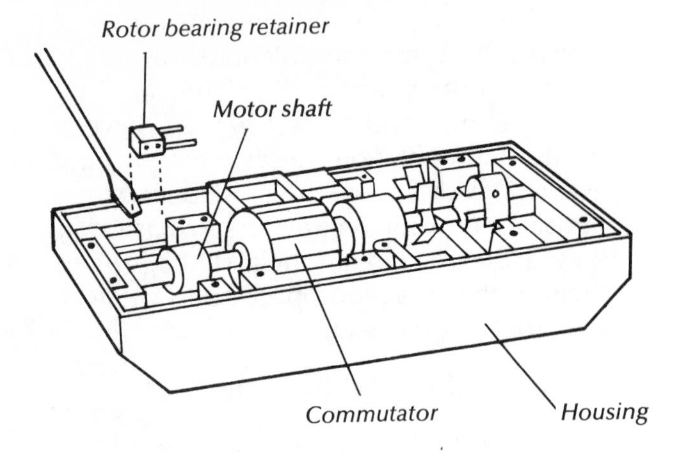


*Inside an electric knife handle. Remove one component at a time; if possible, replace it before going on to the next part*

shape of the part. If you encounter a frozen part, one that will not come loose, apply a few drops of penetrating oil and wait 10–15 minutes, then try loosening it.

You can also try heat when disengaging frozen parts, but use it sparingly. Metal expands when it is warmed, so if you touch the head of a bolt or a rusted nut or screw with the end of a soldering iron, the metal will expand and then contract (as soon as the heat is removed). This action will often free the part and allow it to be disassembled.

Some metal and plastic parts have been put together by force-fitting, and applying heat is the only way you can get them apart. Be careful not to apply so much heat that you melt the parts you are trying to disassemble. Place the point of your soldering iron against the part for a few moments, then remove it, then apply it again after you have tried to loosen it.

## Reassembling

After you have made your repairs in an appliance, it must be reassembled.

1. Check all of the parts in the machine one more time for breakage or wear. Be sure they *all* work mechanically, not just the part you have repaired. When one part of an appliance breaks down, it often causes other parts to wear excessively or to break, so you have to check the entire unit.
2. Clean every part, all connections, everything. If any parts require lubrication, give it to them. Pull gently on all wire contacts to be sure they are tight. Look at the insulation around wires and wrap any cracks or breaks with tape, or replace the wire.
3. Run a continuity test on the entire mechanism (see page 94).
4. Reassemble the appliance in reverse order from the way you took it apart. Here is where all those notes, Polaroid pictures, and diagrams will stand you in good stead. Do not force any parts together. Once they are in place, work them by hand to be sure they are in the correct position.
5. When all of the parts are in place, check them again, then put the outer shell over them and screw its sections together.
6. Plug the appliance in and turn it on. Be alert to any unusual noises, heat, or smells. If anything appears to be wrong, dismantle the appliance and begin checking again.

# MULTIMETERS

The multimeter is a battery-powered test instrument that can function as an ohmmeter, voltmeter, or ammeter. You can use it to determine the continuity of a circuit through a single part or an entire appliance.

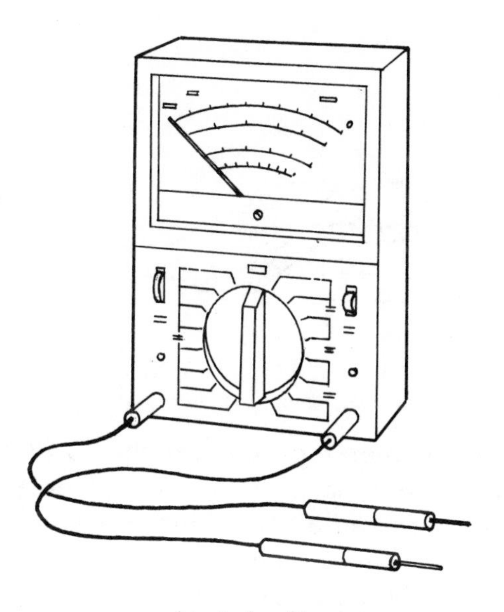

A typical multimeter

You can measure resistance (ohms), voltage, and direct current amperage.

You can buy a multimeter for $200 or more, or spend as little as $15, and the only difference between the two units will be that the $200 unit is a shade more accurate in its readings. Unless you are planning to become a professional electrician, spend the $15.

When you are using any multimeter, remember that it is a battery-powered instrument and supplies all of the current it needs to determine whether there is any resistance in the appliance. **Never test for ohms with the appliance plugged in.** You won't hurt the appliance, but the multimeter will be ruined.

### Using a multimeter

No matter what mode you are in, the multimeter is used in the same manner.

1. Turn the mode switch to the desired function.
2. Touch or clip the meter probes to the test points.
3. Read the correct scale on the face of the unit.

### Multimeter Parts

*Selector switch*—Multimeters have a knob on their face surrounded by several functions that the unit can operate in, including measuring voltage ranges for both AC and DC current, ohmage, and DC amperage.

*Scales*—Above the selector switch is a window with curved scales, which are labeled "Ohms," "VAC" (voltage, AC), "VDC" (voltage, DC), and "ADC" (amperage, DC).

*Pointer*—All of the scales are read by the position of a needle controlled by a zero-adjustment screw. The pointer must rest on zero when the meter is first turned on; you can turn the adjustment screw until the needle is pointing to zero for all of the functions except measuring resistance.

*Ohms-adjustment knob*—Every time you set the meter for reading ohms, the pointer must be adjusted to zero on the ohms scale.

1. Put the test leads in the meter locks.
2. Touch the ends of the probes together.
3. Turn the ohms-adjustment knob until the pointer touches zero on the ohms scale. If the pointer refuses to stop on zero, the meter batteries are bad and need replacement.

*Test leads and jacks*—There are one black and one red test lead and two jacks (holes) for them in the face of the meter. People usually put the black lead in the minus, or common, jack, and the red lead in the plus jack, but the reading will not be affected if you plug the leads in the other way.

*Probes*—The test leads have needle probes or alligator clips attached to them. The alligator-clip probes are particularly handy, since you can clip them to a part and they will stay firmly against the test points.

## Testing for Continuity

The purpose of a multimeter is to help you determine whether electricity is flowing through an electrical component. By testing each component in an appliance, you can verify that the machine is receiving electricity. When testing for continuity, remember that electricity enters any given component through one wire and leaves it through another wire. Consequently, you have to touch the meter probes to both terminals on whatever part you are testing.

1. **Unplug the appliance.** It must not be connected to any power source during a continuity test.
2. Turn the meter's selector switch to the proper ohms position. If you are looking for current leaking from the wiring to the frame of the appliance, use the high end of the ohms scale. Otherwise, always use the low end of the scale.
3. Adjust the pointer to zero ohms with the ohms-adjustment knob. Touch or clip the meter probes together. Bring the pointer to zero ohms by rotating the ohms-adjustment knob.
4. Turn on the appliance. (Don't plug it in, just turn it on.)
5. Touch or clip the meter probes to the prongs of the cord plug. The pointer should move; if it does not, reset the meter to a higher ohms scale. When the pointer moves up the scale you know that both the cord and the On-Off switch are letting electricity into the appliance.

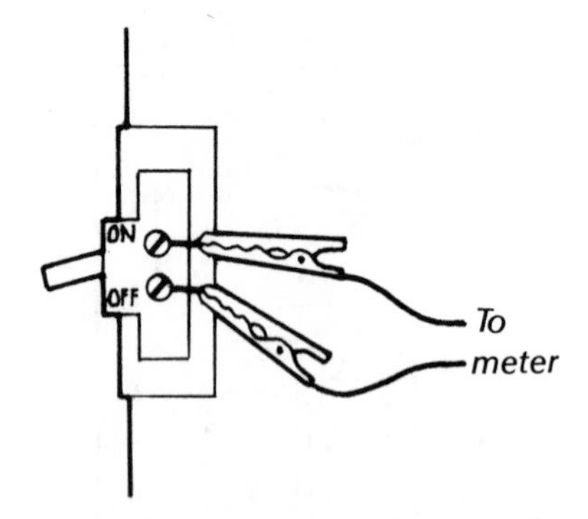


*Touch or clip the probes to the terminals*

6. Put the meter probes on the On-Off-switch terminals. Turn the switch on and off. The needle should jump whenever the switch is on. If it does not move, the switch should be replaced.

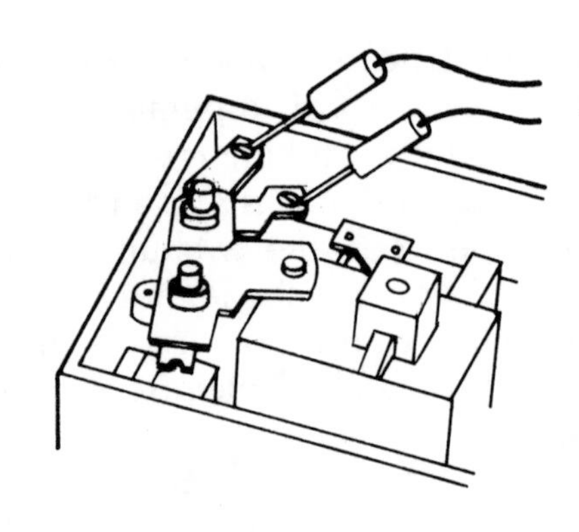

7. Follow the wiring harness to the next part, such as the timer, selector switches, motor, and so on. Any meter reading that is high or "infinite" indicates considerable resistance in the component, and therefore very little current flow. If you get a reading of little or no resistance, there is a short circuit or too much current flow. If you find little or no current passing through a given part, first check for loose or dirty contacts. It is surprising how often a loose screw or speck of dirt can cause an entire unit to stop working.

# WHAT TO DO ABOUT BASIC PARTS

Every appliance has a large number of parts that were made specifically for it, and if anything goes wrong with them, all you can do is write to the manufacturer and ask for a replacement. But there are also a number of components that can be tested and repaired in the same way, even if they look different.

## Cord Sets

A cord set consists of the plug, the cord, and often a strain-relief device in the appliance that keeps the cord from being yanked free of its connections. Cords are whipped, yanked, bent, stepped on, and pulled out of receptacles by their wires instead of the plug. In short, cord sets are badly misused and hence become a common source of trouble. They should be one of the first areas you inspect when the appliance ceases to function properly.

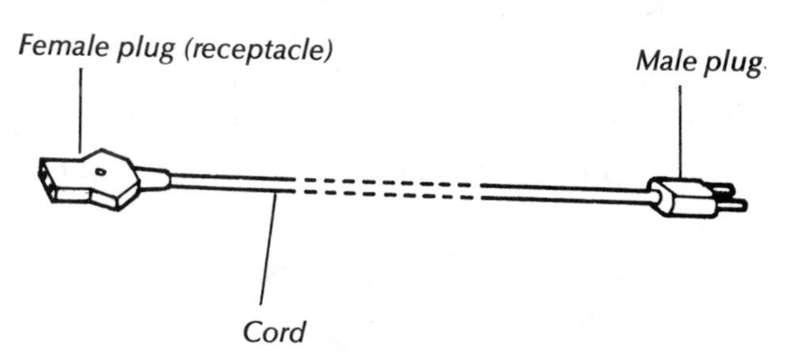


*A cord set consists of a plug (or plugs), cord, terminals, and sometimes a strain-relief device*

Once the cord enters the machine, it will be connected to the timer, the thermostat, the motor, and/or a switch, very often with crimp-on connector terminals squeezed onto the ends of its wires. It is not necessary that you have an exact duplicate to replace a broken cord set, but the wire gauge should be the same or greater. If you are replacing a cord set and you cannot employ the original strain-relief device, you may be able to knot the cord after it enters the appliance and before the wire makes its connections. Use an electrician's knot in the plug (see page 39).

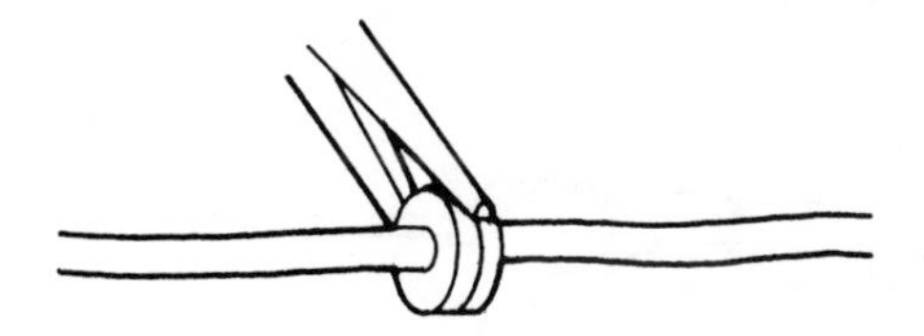

*There are many types of strain-relief devices. Usually they can be taken off the old cord and reused on its replacement*

### *Testing a cord set*

In order to know whether a cord set is permitting current to enter the appliance, or is short-circuiting, it must first be isolated from the appliance circuitry. (Unplug the appliance first, of course.) If the cord is attached to terminals, disengage it and hold the wire ends together with an alligator clip. You can also clip a jumper wire to the wire ends. If you cannot disengage the cord from its terminals (they may be soldered),

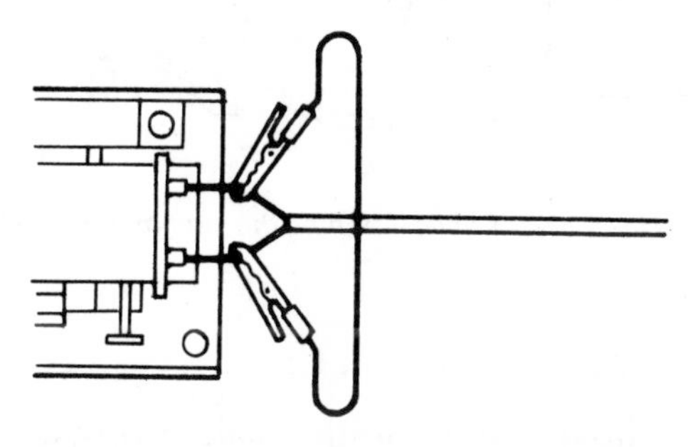

*Use a jumper wire when the cord cannot be disengaged*

clip a jumper wire to the cord wires. If the end of the cord has a female receptacle, either jumper wire clips or meter probes can be inserted in the holes. When the end of the cord has been isolated, clip (or touch) the meter probes to the male prongs of the plug.

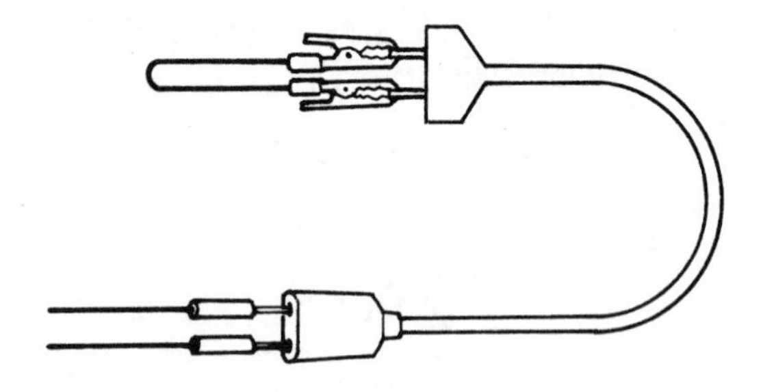

*You can clip a jumper wire to the prongs at one end of the cord and insert the meter probes in the receptacle end*

### *Testing for a short circuit*

1. Set the multimeter to the R×100 scale.
2. Bend and twist the wire and watch the meter needle. If the needle reads zero ohms or jumps when the cord is moved, there is a short circuit in the wire. The cord is OK if the meter reads high.

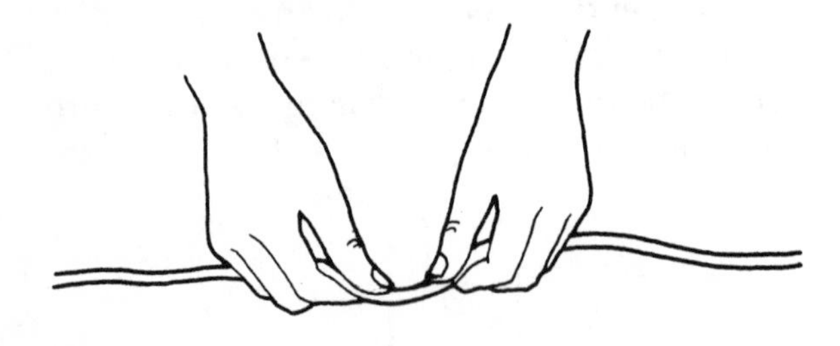

*Flex the cord while watching the multimeter needle*

### *Testing for an Open Circuit*

1. Clip the meter to the male prongs.
2. Clip a jumper cord to the female plug.
3. Set the meter to the R×1 scale.

If the cord is broken, the meter needle will jump or read high. If the cord is OK, the meter will read zero ohms.

## Wiring Harnesses

All of the different parts in an appliance are connected by wires. All of the wires together make up the wiring harness. If the wiring in an appliance (especially a large appliance) has degenerated to the point where it must be replaced, you can purchase a replacement wiring harness from the appliance manufacturer. The harness will arrive in the form of a great many long and short wires, each a different length, each meant to be connected to specific parts, and each color-coded to indicate the appropriate terminals. They will be held together by a plastic ring or two, and have all been precisely cut to run a specific course through the appliance. When you are replacing a wiring harness, do it one wire at a time. Disconnect the old wire of the same color as the new one, and connect the new one to the two open terminals. The new wire should follow the same path as the old.

## Terminals

The terminals found in appliances may be screws or lugs that fit inside crimp-on terminals attached to the wire ends; they may be soldered connections, or have flat or round prongs. Regardless of what the ter-

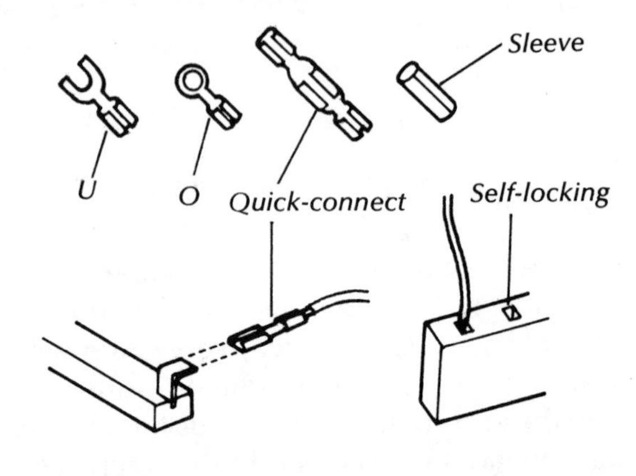


*Some of the many types of appliance terminals*

minals look like or how they are connected, keep them clean to the point of being shiny. You can do this by polishing them with very fine sandpaper or spraying them with a contact cleaner.

## Switches

All switches are designed to break the circuit they are attached to and stop the flow of electricity through them, or close a connection so that current can enter the appliance or a portion of it. Most of the switches you will encounter in appliances are small and cannot be repaired, other than to clean their contacts, but must be replaced if they become defective.

Every switch has both a mechanical and an electrical function, and both must be tested whenever you suspect the switch is faulty. To test the mechanical operation of a switch, push its button, or pull the lever or twist its knob, and observe the smoothness of the action. It should not grind or move unevenly, or bind in any way. If it does, replace the switch. Some switches have metal contact leaves that can be bent slightly so they will meet more readily.

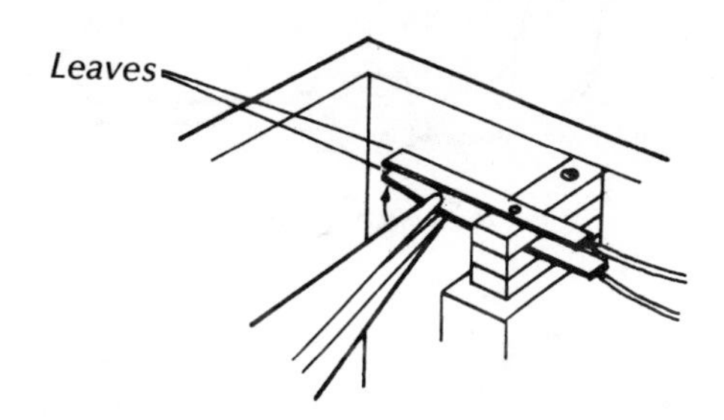


*Bend the switch leaves for better contact*

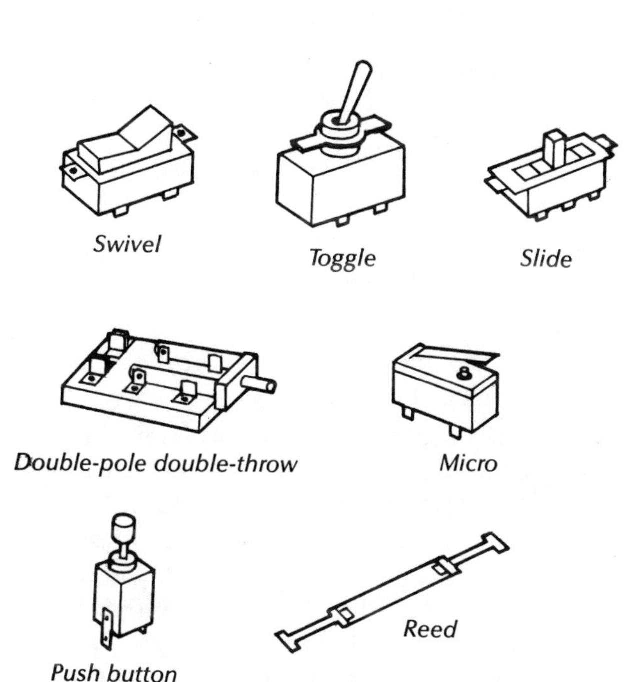


*A few of the types of switches*

Electricity must flow through a switch, entering one side of it and leaving the other, usually through wires connected to terminals affixed to the switch body. If possible, disconnect the switch from the appliance when testing it for continuity.

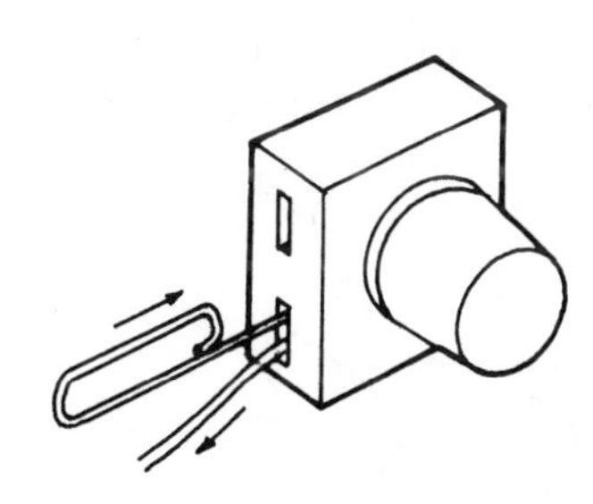

*Use a paper clip to disengage a wire from a self-locking terminal*

### *Testing switches for continuity*

1. Clip the meter probes to the switch contacts.

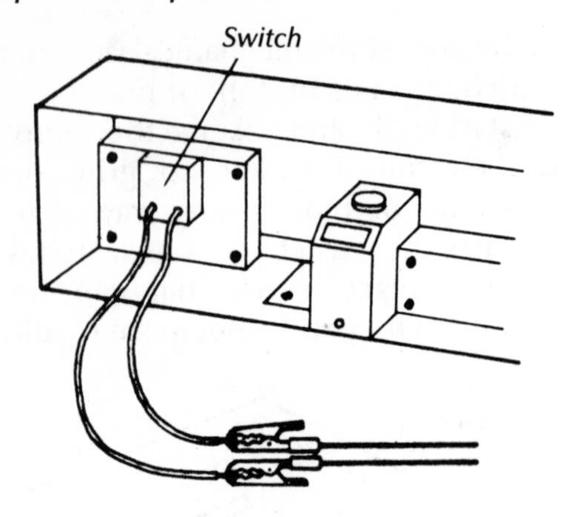


*Test the switch contacts*

2. Set the meter to the R×1 scale.
3. Turn the switch on and off several times.

The switch is electrically functional if the meter reads between zero and 5–12 ohms.

## Timers

It can be argued that the ultimate switch is a timer, which is the essential control center of any appliance that operates in cycles (clothes washer or dryer, dishwasher). Timers all look a little different, but they function in about the same way. They are operated by a small synchronous clock motor that turns a series of cams on a shaft. As the cams move around the shaft, they touch a series of electrical contacts at different parts of the machine; consequently, there may be as many as 20 colored wires leading away from the timer, which go off to activate a motor, solenoid valves, switches, and so on. The front of the timer almost always has a dial that you rotate to set the timer for performing the various functions of the appliance.

If an appliance stops operating properly, or does not run at all, start your diagnosis of the problem at the timer.

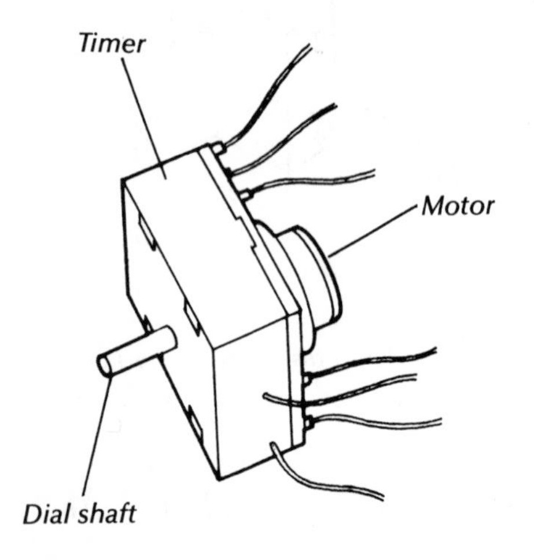


*Timers can have as few as two wires or more than twenty*

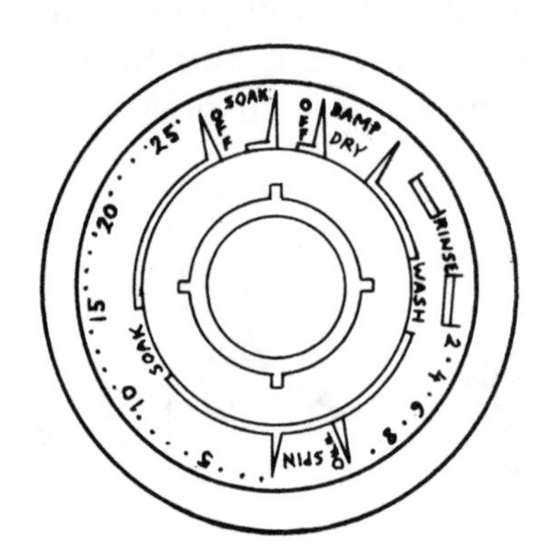


*Washing machine timer dial*

1. Open the appliance so that you can see the back of the timer.
2. Turn on the appliance and observe the timer in action. Specifically, look at the motor attached to its side. You will be able to see the edges of several small gears, which all should be turning at different rates. If they are not rotating, the motor is probably faulty. You can try to get the gears moving again with two or three drops of light machine oil, but most likely the motor will have to be replaced.

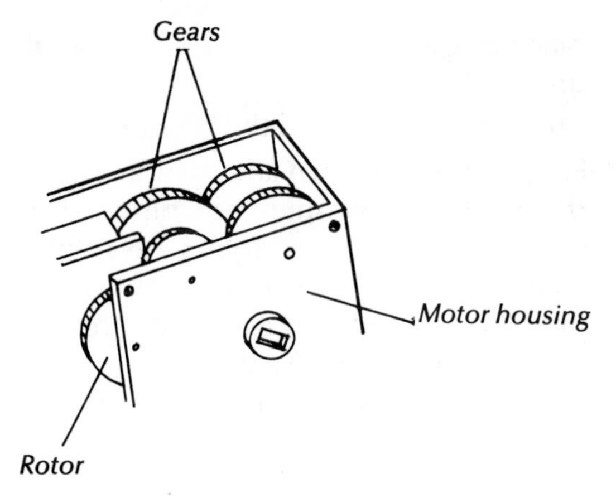


*Watch the timer motor in action*

Timer motors are tiny and you need a set of jeweler's tools to repair them, so nobody bothers. To replace a timer motor, disengage the two wires (one black, one white) that go from the motor to the back of the timer. Undo the screws or bolts that hold the motor in place and remove it. Attach the replacement motor to the timer and connect the two wires to the terminals used by the old motor.

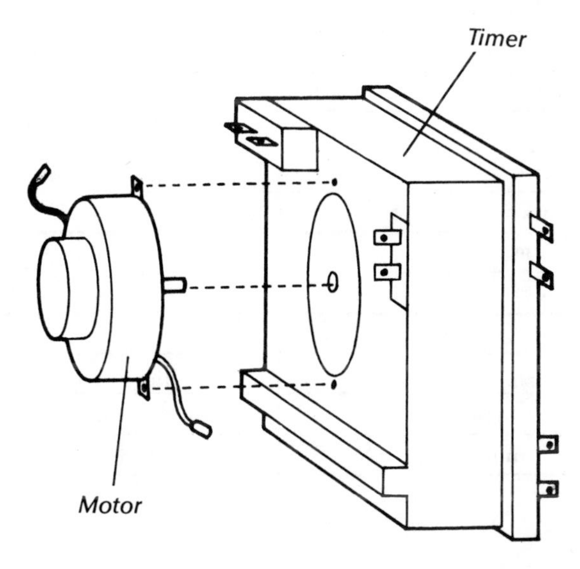


*Only two wires go from the motor to the timer*

3. If the motor is functioning, the problem is in the timer itself, and repairing it requires the services of special equipment and experience. The cams may be worn, or the contacts may be dirty. Worn cams cannot be repaired, but dirty contacts can be cleaned. If cleaning the contacts fails to get the unit working again, you could test the timer for continuity. However, all 18 or 20 of its wires must be tested, and when you finish, all you will know is which ones are not working. And the unit will still have to be replaced.

### *Replacing a timer*

1. **Do not remove the old timer from the appliance.** Write down the parts number embossed on its casing and take it, together with the appliance model number and manufacturer's name, to a replacement-parts store (or write to the manufacturer). The parts store will sell you a replacement timer. You will have to pay a deposit ($10–$20), which will be returned when you bring in your old timer.
2. **Unplug the appliance or turn off the branch circuit.**
3. Remove the timer dial from the front of its shaft. This is usually just pulled off its shaft, but there may be a setscrew to loosen.
4. Remove one wire from the old timer and install it on its corresponding terminal on the new unit. The wires are all different colors, and they and their terminals may also be labeled with corresponding letters or numbers.
5. Continue to transfer one wire at a time from the old timer to the new timer. Be aware that you may make a mistake during the transfer of wires, with the result

that the new timer does weird things to the appliance when you run it. You will not destroy the appliance or the timer by wiring it incorrectly.

6. When the old timer is completely disconnected electrically, you can unbolt it from the appliance and remove it, but first carefully note its position. For example, it may be tipped to one side.
7. Bolt the replacement timer in position, placing it at the same angle as the old unit.
8. Attach the face dial to the timer shaft.
9. Plug in the appliance and turn it on. The cycles of most washing machines, dishwashers, and clothes dryers are printed somewhere in the owner's manual. The printout will tell you how long the machine washes, rinses, pauses, or whatever, through each of its cycles. Set the

CYCLE DESCRIPTION

| REGULAR | | PERMANENT PRESS | | DELICATE | |
|---|---|---|---|---|---|
| Fill | Time variable | Fill | Time variable | Fill | Time variable |
| WASH | Select. to 10 min. | WASH | Select. to 10 min. | AGITATION | 2 minutes |
| Pause | 1 minute | Pause | 1 minute | SOAK | 3 minutes |
| SPIN & DRAIN | 2 minutes | SPIN & PAR. DRAIN | 45 sec. (Approx.) | AGITATION | 1 minute |
| SPRAY RINSE | 1 minute | Fill with cold water | Time variable | SOAK | 3 minutes |
| SPIN & DRAIN | 1 minute | SPIN & PAR. DRAIN | 15 sec. (Approx.) | AGITATION | 1 minute |
| Fill | Time variable | Fill with cold water | Time variable | SOAK | 1 minute |
| DEEP RINSE | 2 minutes | Pause | 1 minute | SPIN & DRAIN | 2 minutes |
| Pause | 1 minute | SPIN & DRAIN | 45 sec. (Approx.) | SPRAY RINSE | 1 minute |
| DAMP DRY | 5 minutes | SPRAY RINSE | 15 sec. (Approx.) | SPIN & DRAIN | 1 minute |
| | | SPIN & DRAIN | 1 minute | Fill | Time variable |
| | | Fill with cold water | Time variable | DEEP RINSE | 2 minutes |
| | | DEEP RINSE | 2 minutes | Pause | 1 minute |
| | | Pause | 1 minute | DAMP DRY | 5 minutes |
| | | DAMP DRY | 3 minutes | | |
| | | Timer moves "OFF" | 2 minutes | | |

| TIMED SOAK | | SOAK ONLY | |
|---|---|---|---|
| Fill | Time variable | Fill | Time variable |
| AGITATION | 2 minutes | AGITATION | 2 minutes |
| SOAK | 7 minutes | Timer moves "OFF" | 1 minute |
| AGITATION | 1 minute | | |
| SOAK | 6 minutes | | |
| AGITATION | 1 minute | | |
| SOAK | 7 minutes | | |
| AGITATION | 2 minutes | | |
| SOAK | 1 minute | | |
| SPIN & DRAIN | 2 minutes | | |

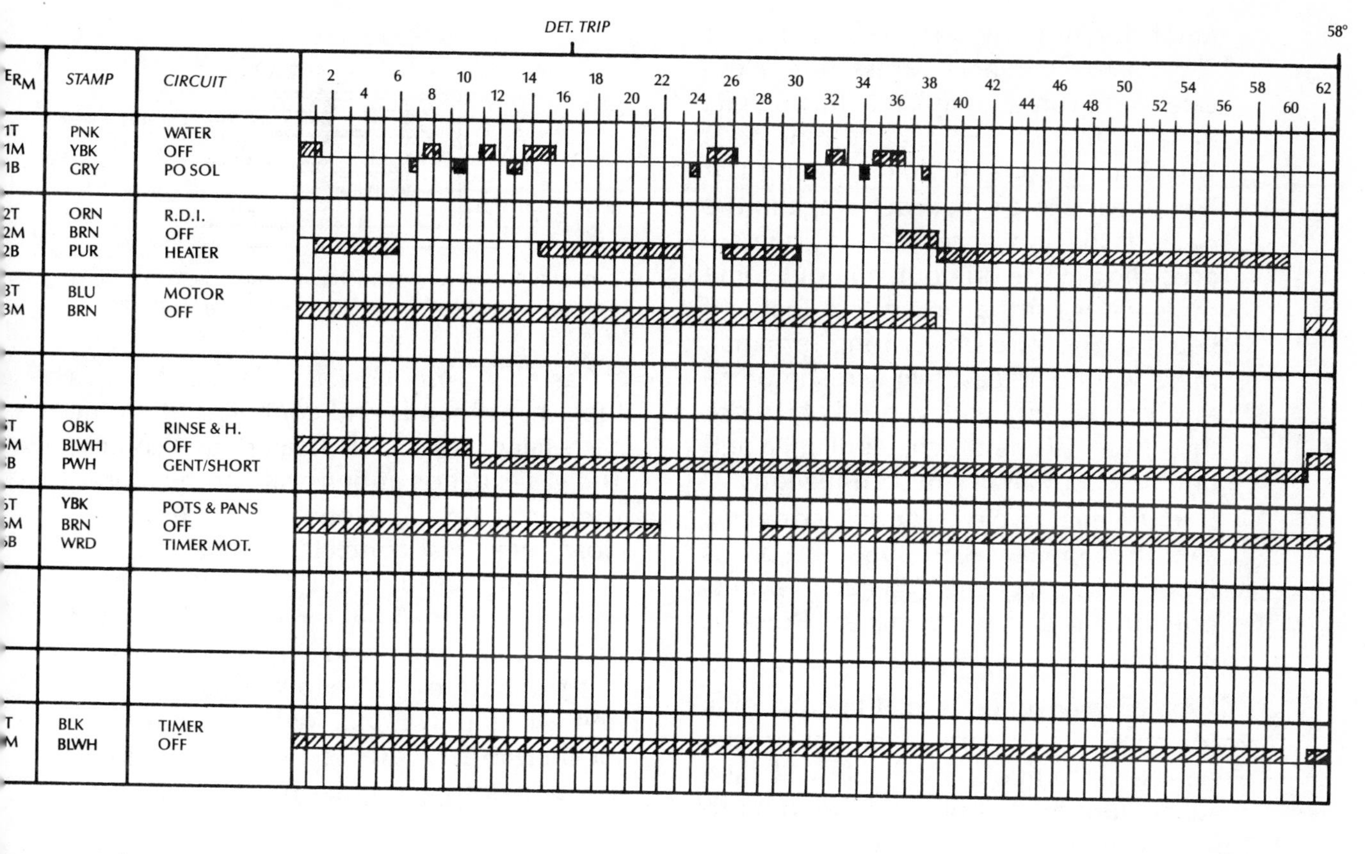


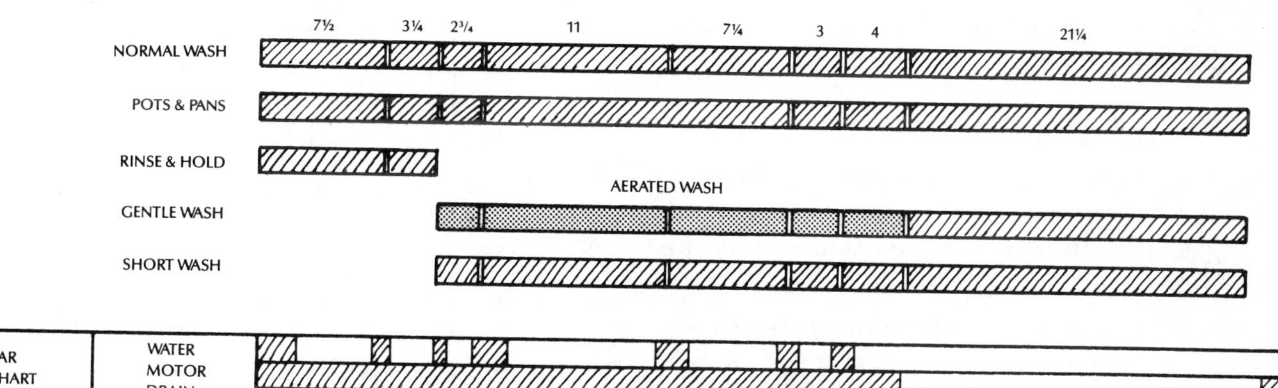


*This timer information might be printed in the owner's manual for a washing machine*

*This is the kind of timer chart you might find pasted on the back of a dishwasher—everything you need to know about the timer*

SWITCHES

| 1-14 | 5-14 | 9-14 | 13-14 |
|---|---|---|---|
| X | X | O | O |
| O | X | O | O |
| O | O | O | X |
| X | O | X | O |
| X | O | X | O |

O—OPEN
X—CLOSED

timer for a particular cycle (such as "Regular," "Permanent Press," or "Delicate") and run it completely through its cycle, noting whether the machine does what it is supposed to do for as long as it is supposed to do it at each stage in the cycle. If the appliance does not operate according to the printout, you put some of its wires on the wrong terminals. (If you put one wire on a wrong terminal, there will be at least two wires improperly connected, and very likely more than two.) Dismantle the appliance and double-check all of your connections.

If the first cycle performs correctly, run the appliance through all of its modes. The fact that "Regular" works does not mean that "Permanent Press" is also correct.

If you have never replaced the timer on a major appliance before, and you are an electrical nincompoop, the job will take about an hour to do—plus another hour to correct your mistakes. But you will save roughly $60 in repair bills, and feel very pleased with yourself in the bargain.

## Thermostats

Appliances have three types of thermostats: bimetal strips, thermodisks, and gas-filled units, which are usually found in ovens and ranges.

*Bimetal Strips and thermodisks*—These are small pieces of metal composed of two or more alloys bonded together. When heated, the metals expand and contract at different rates, which makes them bend toward or away from a stationary electrical contact. If a thermodisk or bimetal strip does not operate properly, it can only be replaced with an identical component.

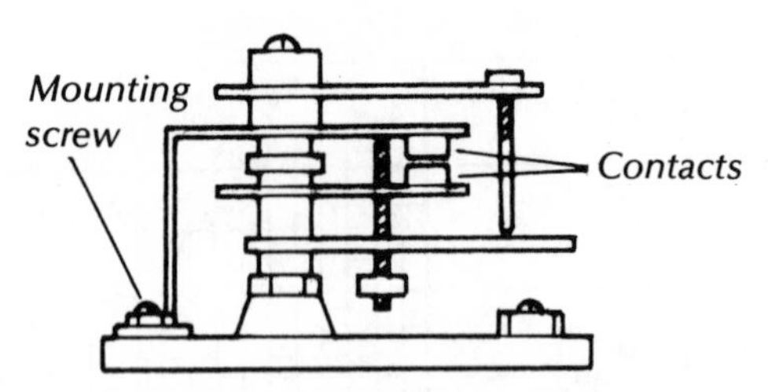


*Surrounded by other parts, bimetal strips and thermodisks may appear formidable, but they are easily removed for replacement*

*Gas-filled thermostats*—These have a tube that expands and contracts when the gas sealed inside it is heated and cooled. The gas expands under heat and puts pressure on a small bellows, which opens or closes electrical contacts. By changing the dial setting on the front of the thermostat you can control how long it will take the heated gas to open and close its bellows.

If the tube or its bellows leaks gas, the unit will stop working. Rather than replace the thermostat, have it rebuilt—it costs half as much as buying a new thermostat.

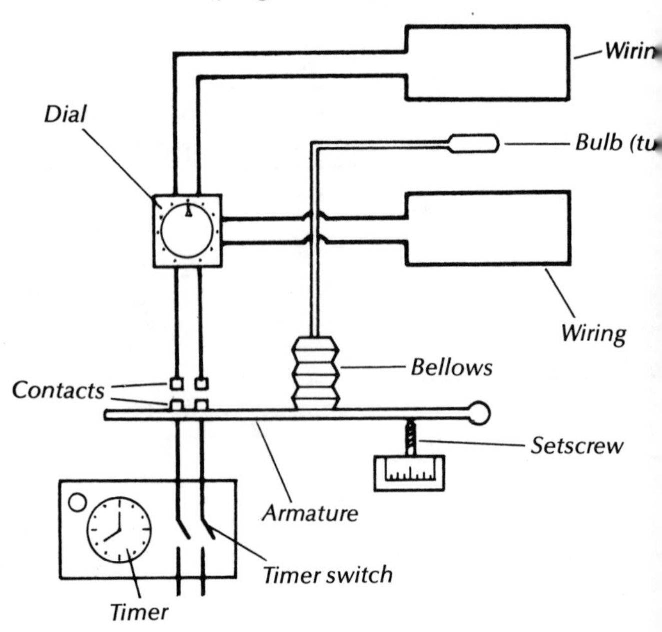


*Gas-filled thermostat*

### *Testing thermostats*

1. **Turn off the power to the appliance.**
2. Set the multimeter to the R×1 scale.
3. Turn the appliance to its highest (hottest) setting.
4. Touch the meter probes to the terminals on either side of the thermostat. If the meter reads high, the thermostat is bad and should be replaced. If the meter reads zero ohms, the thermostat is OK.

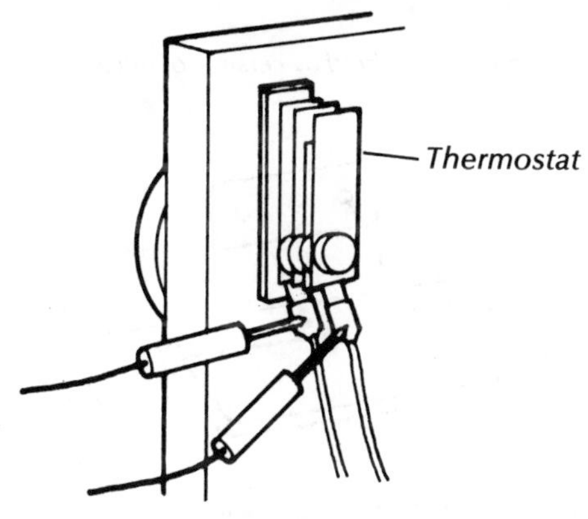


*Testing a thermostat*

### *Replacing thermostats*

Replace a bimetal strip or thermodisk by disconnecting its electrical contacts and undoing the screw or bolt that holds it in position. Secure the new unit in place and connect the wires to their terminals.

A gas-filled thermostat may have more than one wire attached to it, so be careful to note where each wire is connected before you remove it from the old unit. When replacing a gas-filled thermostat, you must install not only a new bellows but the sensing tube and bulb as well. Disengage all parts of the thermostat and lift the entire unit out of the appliance.

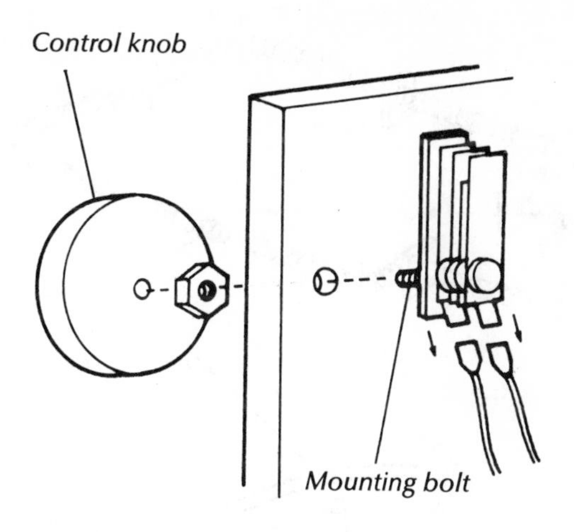


*Thermostats are easily removed and replaced—usually a matter of a couple of contacts, a few nuts or screws, and perhaps a knob to disengage*

## Heating Elements

Heating elements in toasters and other small appliances often take the form of a wire whose resistance to the flow of electricity causes the wire to heat up. In ovens, ranges and cook tops, as well as dishwashers and other appliances, the heating element is often a solid rod, or wires embedded in a ceramic casing.

Heating elements should never be repaired, since any repair creates a weak link in the element and diminishes its performance. If the element is a coiled or straight wire, it was strung around a series of ceramic holders, with its ends connected to electrical wires. The bar or rod elements usually plug into what amounts to an electrical socket and are simply unplugged for replacement. The replacement of heating elements in a large appliance is always described in the service manual and in many cases also in the owner's manual.

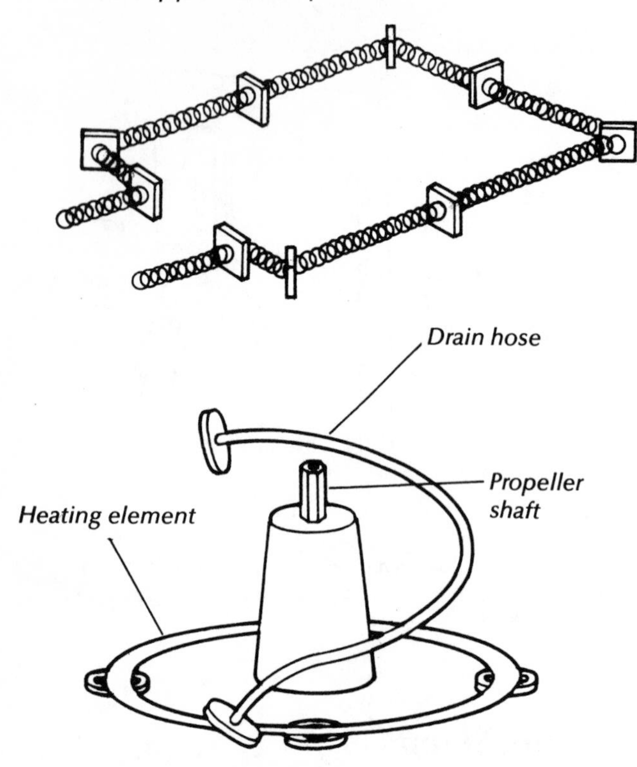


*Heating elements can be coiled resistance wires (above, as in a toaster) or solid metal rods (below, in a dishwasher)*

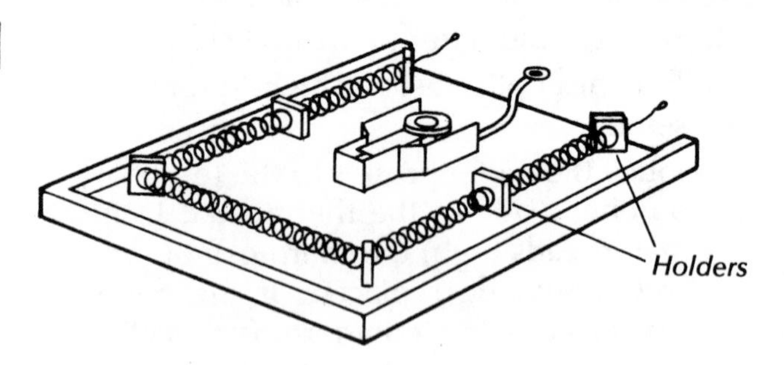


*The resistance wire is held by ceramic or mica holders*

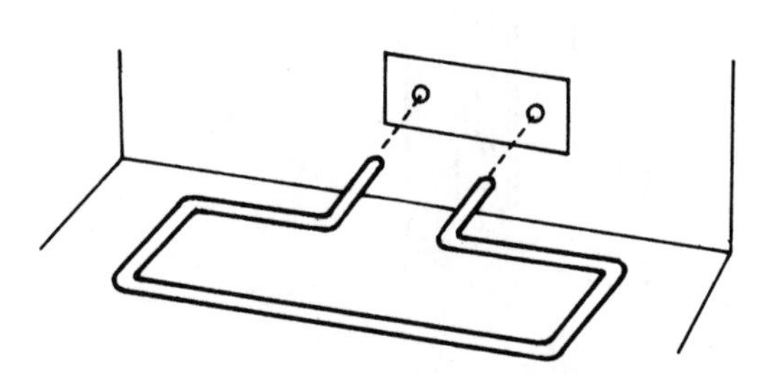

*Rod elements plug into a socket*

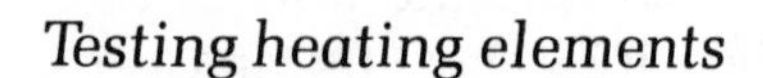

### Testing heating elements

1. **Unplug the appliance.**
2. Remove the heating element from the appliance circuit.
3. Set the multimeter to the R×1 scale.

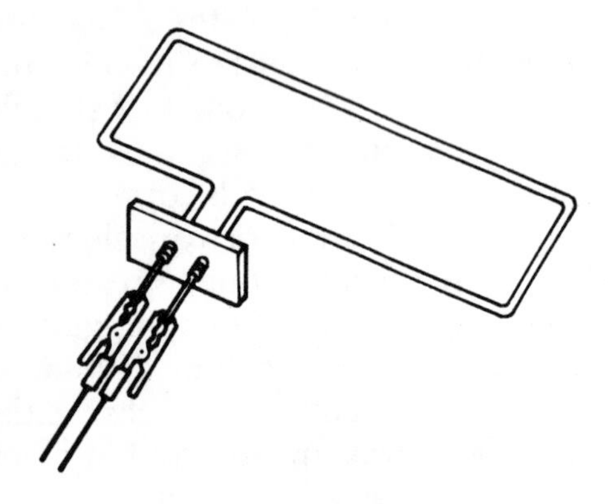

*Testing a heating element*

4. Clip the meter probes to the heating element terminals. If the meter needle reads 10–20 ohms, the element is OK. If the meter reads high, the element is dysfunctional and must be replaced.

## Solenoids

Solenoids open and close switches and water intake and outflow ports. A solenoid is merely a coil of wire wrapped around a tube. When current flows through the wire, it creates a magnetic field that attracts or repels a plunger in the tube. The plunger then opens or closes the port or trips a switch.

The electrical terminals can corrode. Clean them, and lubricate the plunger if it does not move smoothly. If the plunger is jammed, first clean it; if it is bent, it must be

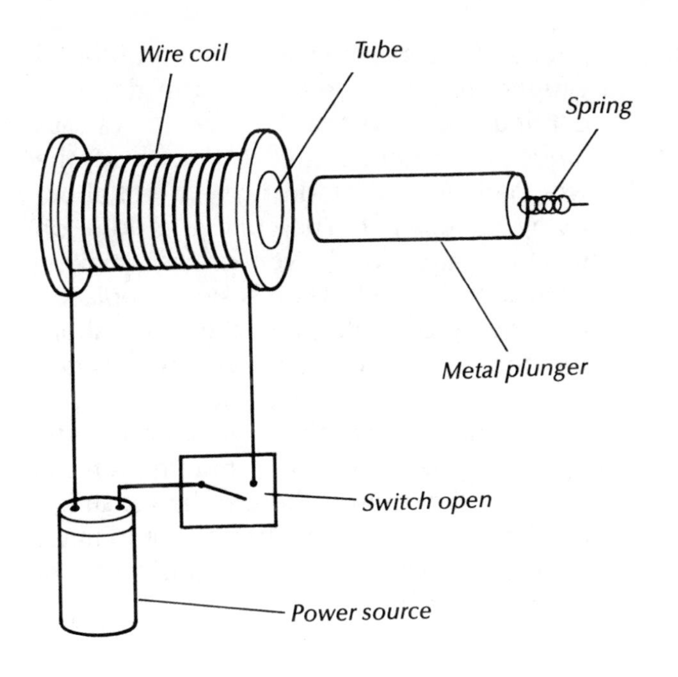


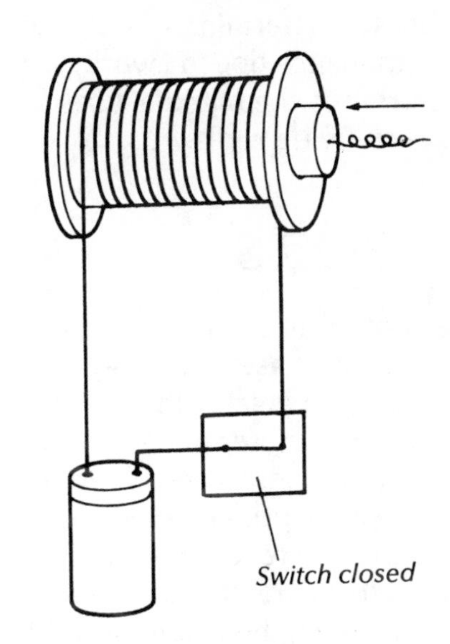


How a solenoid works

replaced. If the return spring is worn or broken, replace it. If there is no continuity in the solenoid, or you cannot make it work smoothly, replace the entire solenoid.

*Testing a solenoid*

1. **Unplug the appliance and disassemble it.**
2. Disconnect one wire attached to the solenoid.
3. Set the meter to the R×100 scale.
4. Clip the meter probes to the solenoid terminals. If the meter reads high, the solenoid is faulty. If the meter reads 250–1,000 ohms, the solenoid is OK.

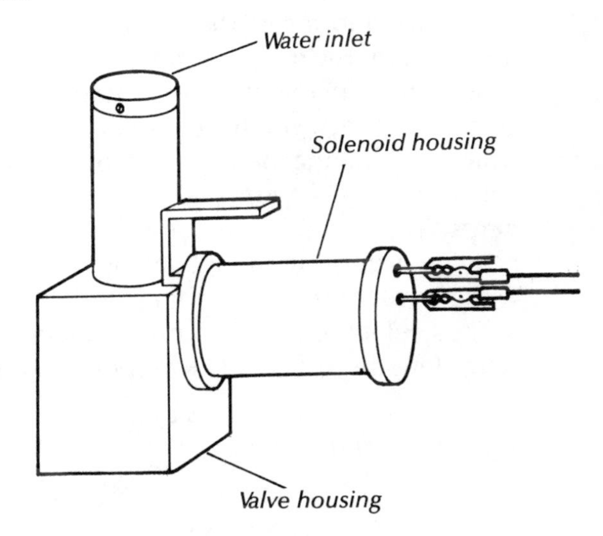


The solenoid being tested controls a water valve

## Solid-State Controls

Most large and small appliances sold today incorporate partial or complete solid-state control systems. The printed circuitry is such that it cannot be repaired, so the solid-state part of the appliance is usually covered by a long-term manufacturer's guarantee to replace it should anything go wrong. You can test a solid-state component for continuity by setting your meter to the R×1 scale and touching the probes to the terminals on the circuitry, but if the appliance doesn't work, all you can do is return it to the manufacturer and ask that the circuitry be replaced.

# MOTORS

The motors used in home appliances are either split-phase, shaded-pole, capacitor-start, or Universal. Although they all look somewhat different, they have the same basic components and operate according to the same principle: Electricity flowing through a wire creates a magnetic field around that wire. If you coil the wire around a hollow core, the magnetic field is concentrated in the center of the hole. So if you coil a second wire around a solid core, such as a shaft, and insert the solid core in the hollow core, the magnetic field will spin the solid core. Once you have a rotating solid core (armature), you have a spinning shaft, which can be attached to a fan or gears or any number of work-producing components. The secret to keeping the motor in running order is to not let any of the coiled wires break or touch anything.

Thus, electric motors consist mainly of long lengths of wire wrapped around pieces of metal. The wires can be different gauges and must be wrapped in different configurations so that they create various kinds of magnetic fields. If one of the wrapped wires should break, forget about trying to repair it. The coiling is done on machines and must be very precise, making it cheaper and a whole lot easier to replace a defective motor altogether. In fact, about the only repairs you can make to motors is to clean the electrical contacts, lubricate them, and tighten the mechanical parts. Continuity testing with a multimeter will help you determine whether the repair is a mechanical one you might be able to make or an electrical one beyond the scope of your equipment.

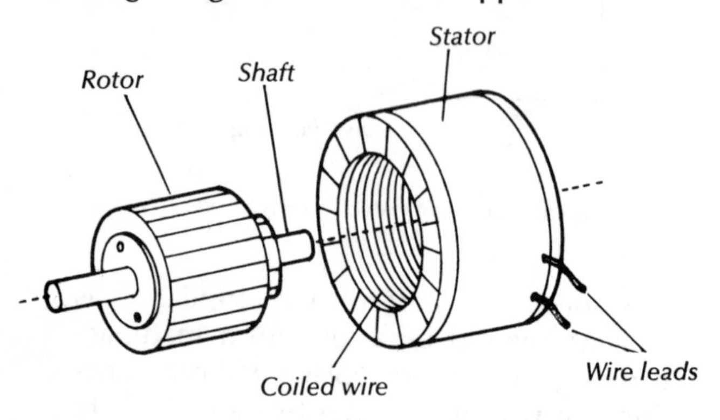


*Components of an electric motor*

## *Split-Phase Motors*

These are used in fans, washing machines, oil burners, pumps—any appliance that requires single-speed, heavy-duty power. Along with their rotor, a main winding, start winding stator, and end plates, they may also have a centrifugal starting switch.

## *Shaded-Pole Motors*

Shaded-pole motors are used in fans, blowers, hair dryers, and other appliances that have low starting torque. They are reliable and consist of a stator (field frame), a rotor, and end plates.

## *Capacitor Motors*

Capacitor motors are used primarily in refrigerators, freezers, air conditioners, oil burners, washing machines, and pumps, or any appliance that needs an unusually powerful motor. There are several types of capacitor motors, but they all work the same as the split-phase motor.

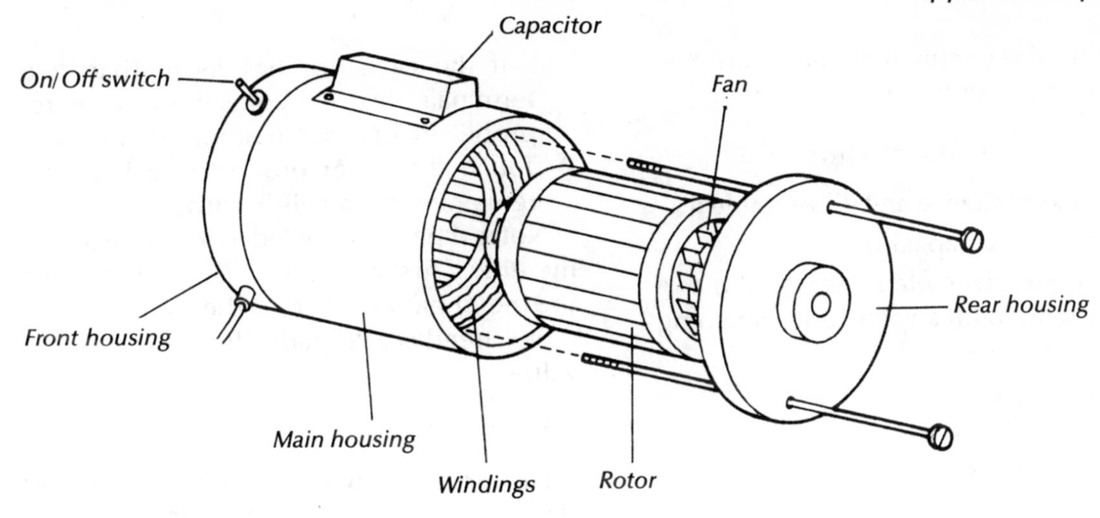


Split-phase motor

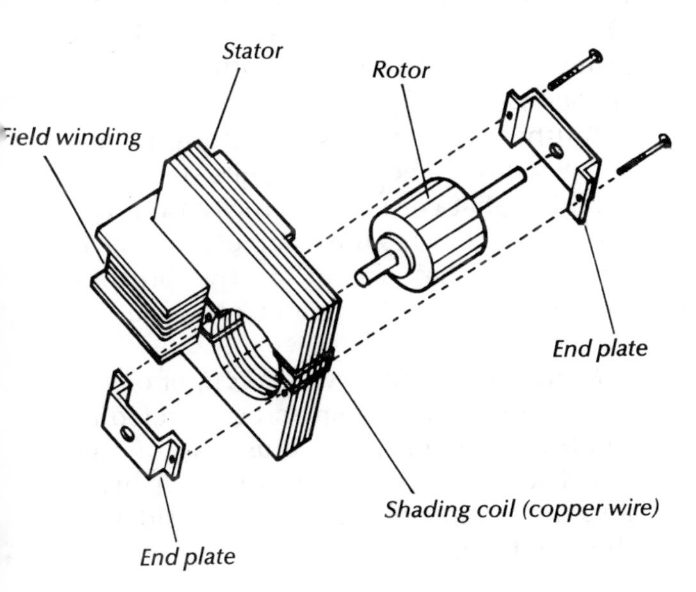


Shaded-pole motor

*Capacitors*—A capacitor is a storage unit for electricity. Connected to a motor, it provides a slight shift in the timing, or phase, of the electricity entering the motor's start winding and provides an added boost of current to the starting torque of the motor. There are two kinds of capacitors: electrolytic and oil-filled.

*Electrolytic capacitors*—These are two sheets of aluminum foil separated by gauze soaked in a chemical solution called electrolyte; the foil and gauze are sealed inside an aluminum or plastic canister. Electrolytic capacitors are supposed to function for only a few seconds at a time, and are disengaged from the circuit as soon as the motor has reached 75% of its operating speed.

*Oil-filled capacitors*—These are sealed containers filled with oil-saturated paper. There are also two sheets of foil connected to terminals on one end of the housing. Oil-filled capacitors are able to perform continuously and do not have to be re-

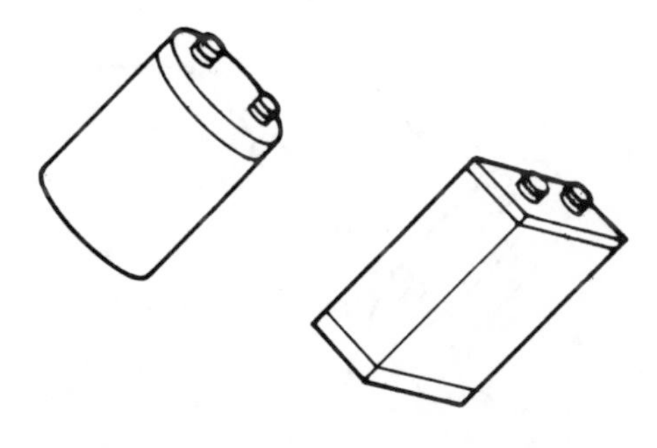

Capacitors can be round or rectangular

moved from the circuit after the motor has reached its running speed.

### *Testing capacitors for resistance*

1. **Unplug the appliance and disassemble it.**
2. Disconnect the capacitor.
3. Lay a screwdriver blade across the terminals to drain off any current remaining in the unit.

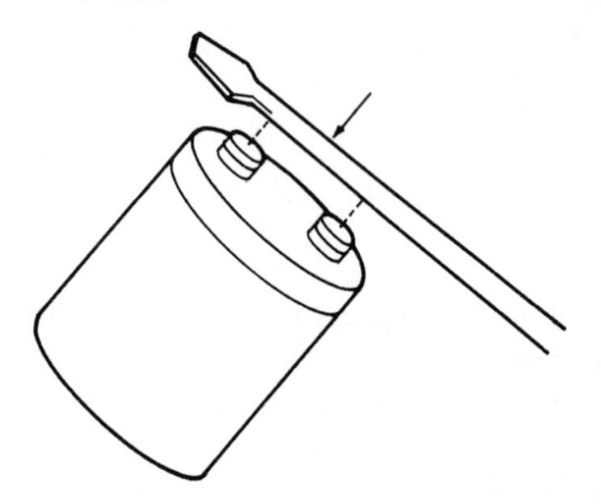

*Use a rubber-handled screwdriver to drain off resistance*

4. There is an ohms rating printed on the casing of the capacitor. Set the multimeter according to the printed rating.
5. Touch a meter probe to each terminal on the capacitor. If the reading is less than 10 ohms, the capacitor is shorted out. Replace it with a unit having the same ohms rating.

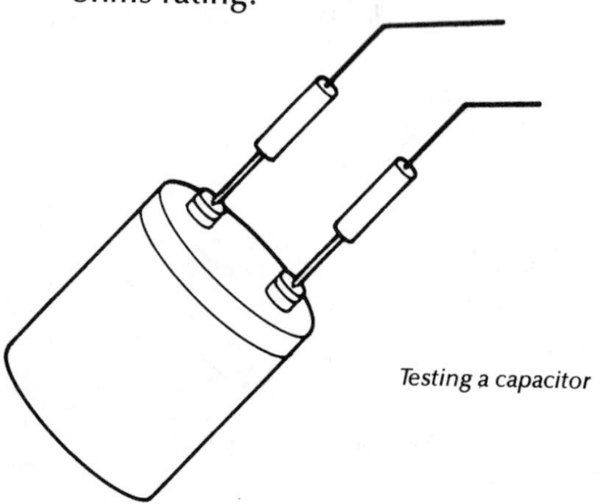

*Testing a capacitor*

If the capacitor has more than two terminals, test it by placing one meter probe on the "C" (common) terminal. Touch the other probe to each of the remaining terminals in turn.

**Note:** When connected to a capacitor, the meter will first read low. As electricity from the meter battery charges the capacitor, the meter needle will gradually reach a higher value.

## *Universal Motors*

Universal motors are found in vacuum cleaners, small power tools, blenders, and sewing machines; they are small, very reliable power plants that work on either AC or DC current. Universal motors consist of a frame, field core, armature, brushes, commutator, and end plates.

Brushes are the unusual feature of Universal motors. They are graphite or carbon cylinders that conduct electricity from the motor's field coil to the commutator, which causes the shaft to rotate. The brushes eventually wear down and must be replaced.

Many brushes come with a copper pigtail wire embedded in them and connected to the brush holder. If the brush is not pigtailed, it is inserted in the brush holder with a spring that pushes it against the sides of the commutator.

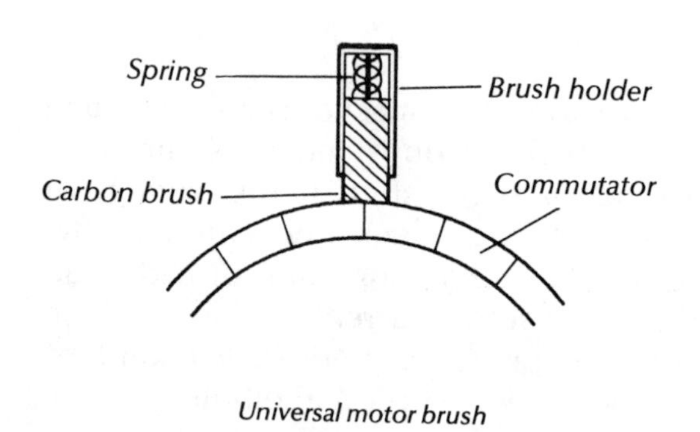


*Universal motor brush*

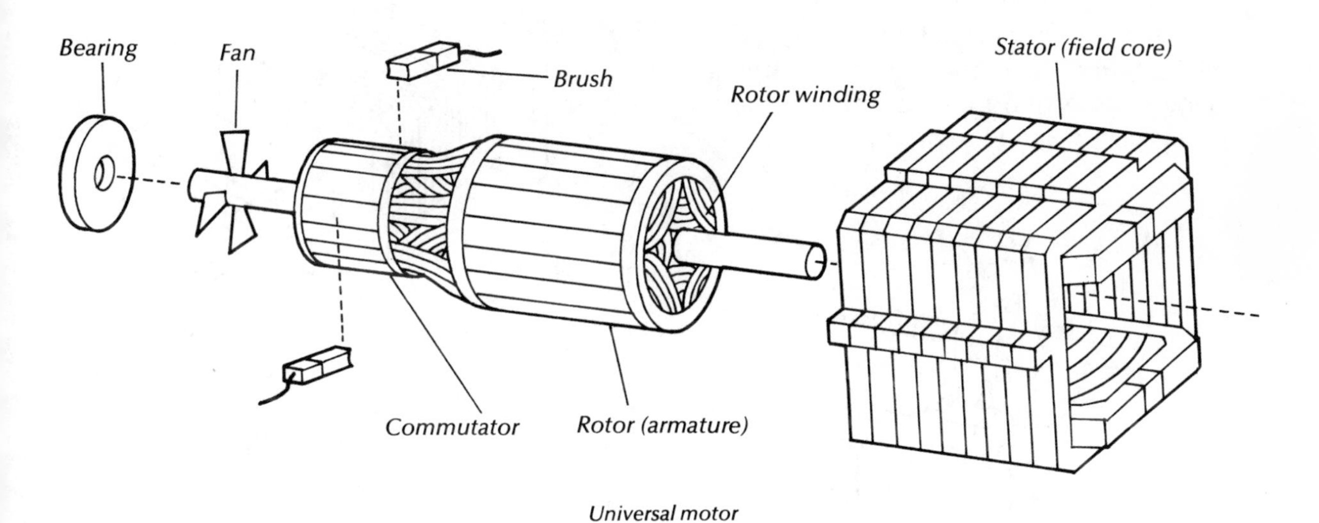


Universal motor

## Troubleshooting Motors

When a motor stops functioning properly, troubleshoot it to determine whether it needs new bearings, leads (insulated wires), or switches, and whether the windings have burned out. In most cases you can test a motor without removing it from its appliance, but **always unplug the appliance when testing for resistance (ohms).**

1. Check the cord and motor end plates for cracks or breaks. Be sure the shaft is not bent and that leads are not broken or burned.

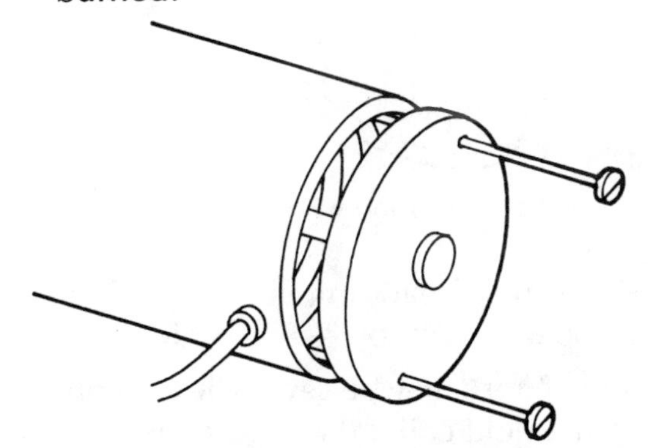

Check the cord and end plates

Unscrew the brush holders on a Universal motor and remove the brushes. If the brushes are wider than they are long, replace them. Check the brush springs for wear and breakage. Brushes must

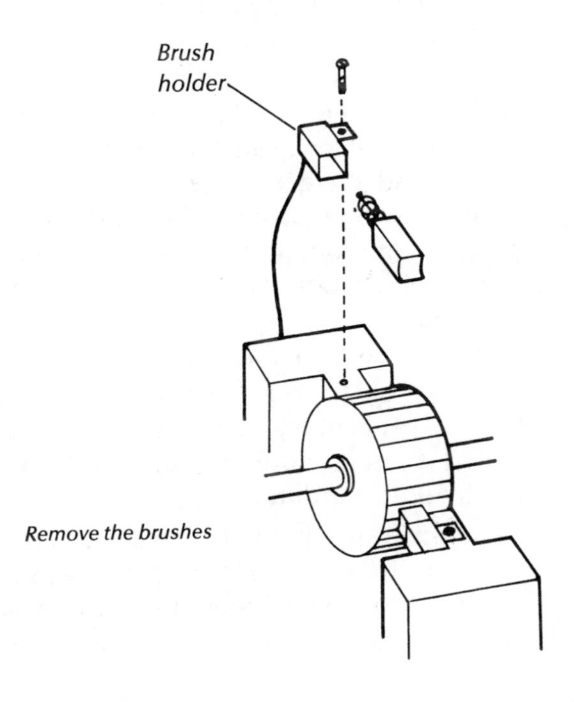


Remove the brushes

move freely; their springs should push them firmly against the commutator.

New brushes have flat ends. Curve the ends to fit the commutator by wrapping sandpaper around the commutator and rotating it to sand the brushes.

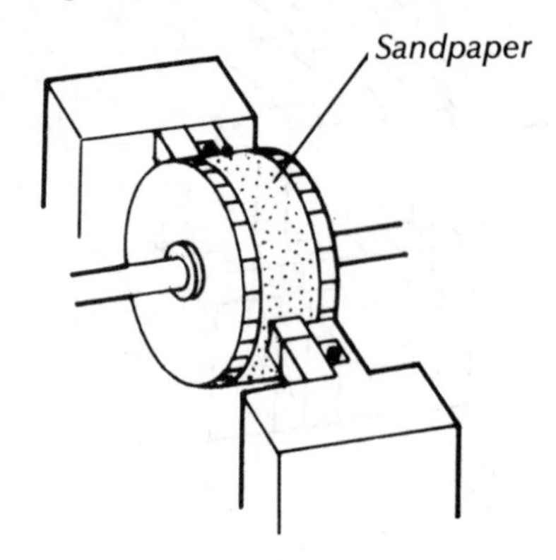


*Rotate the commutator by hand*

2. Jiggle the motor shaft. If it moves, the bearings are worn. Spin the shaft. If the shaft does not rotate freely, it could be bent or the bearings could be bad. Either situation can cause a circuit interrupter to blow or trip.
3. Turn on the motor. Place one probe of a hot-line tester against the running winding lead (larger-gauge wire). Put the other probe against the core. If the light goes on, the winding is grounded. Look for any wires touching the rotor or stator cores.
4. Run the motor for a few seconds. If it blows or trips the circuit interrupter, smokes, is noisy, runs too slowly or too fast, there is a burned or partially short-circuited winding.
5. Remove the motor end plates. Pull the stator from the rotor. If the winding wires are black, they are burned and must be replaced.

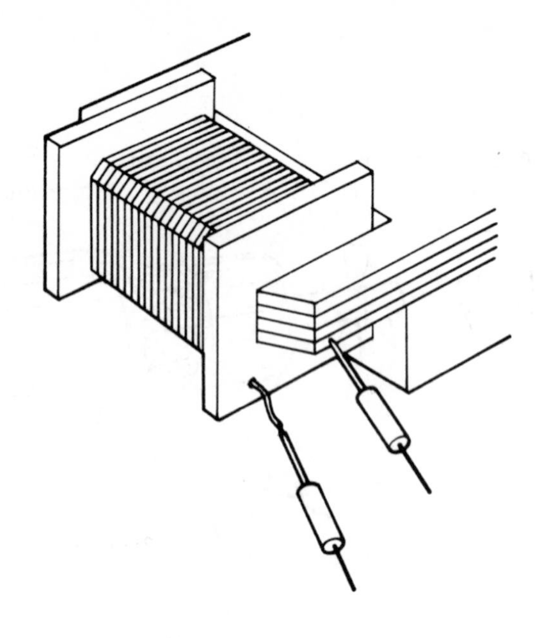

*Testing a field coil*

## Testing field coils for open circuits

1. Set the meter to the R×100 scale.
2. Touch a meter probe to each of the field coil leads.

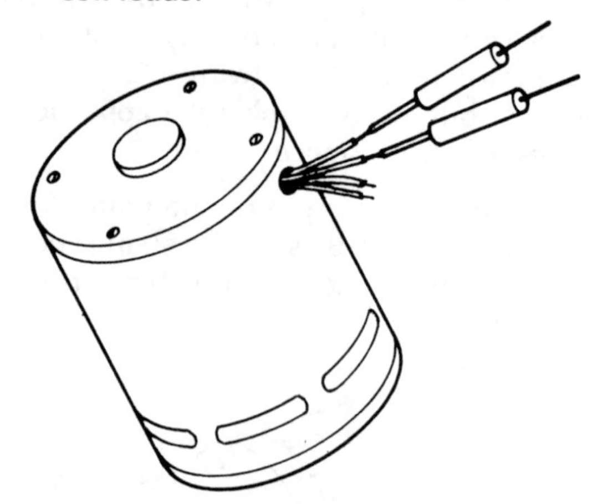

*Testing for open circuits*

3. If the meter reads infinite ohms, there is an open circuit (broken wire) in the coil.

**Note:** There may be several leads coming from the field coil if the motor has variable speeds. Each lead must be tested for resistance.

### *Testing field coils for leakage*

1. Set the meter to the R×100 scale.
2. Touch one probe to the field coil lead.
3. Touch the other probe to the motor frame.

   If the meter reads high, current is leaking to the motor frame and probably causing a shock. If the meter reads infinite ohms, there is no leakage.

**Note:** You can also test the armature for leakage by touching the meter probes to the motor shaft and the commutator bar.

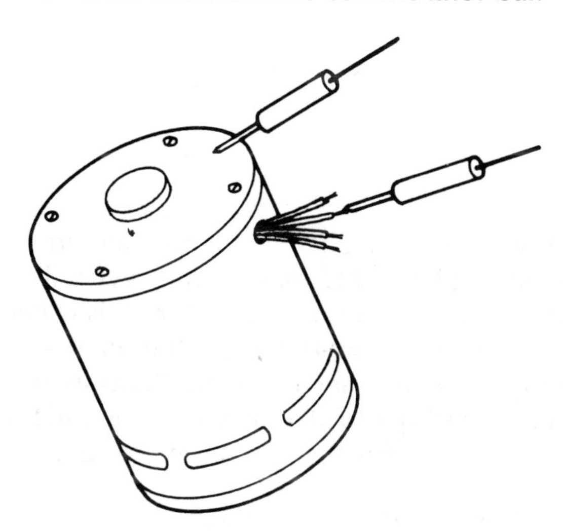

*Testing for leakage*

### *Special tests for Universal motors*

1. Inspect the fan blades on the shaft of the Universal motor. They should not be bent or broken; the fan must be securely on the rotor shaft.
2. Look at the commutator. Its brass bars are divided by mica strips, which are not supposed to be higher than the surface of the bars. If the mica strips protrude,

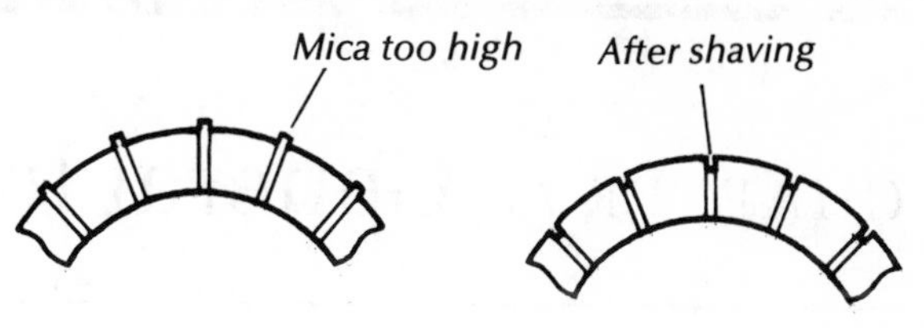


*Shave any mica strips that are higher than the commutator bars*

shave them down with a knife to below the level of the brass. If the brass bars are pitted, very shiny, or discolored, there is a short circuit in the armature coil. If the bars are rough, sand them smooth with very fine sandpaper. The roughness comes from defective brushes, so also replace the brushes.

*Testing Universal motor commutators*—If the motor runs hot, sparks, or does not start, there is a short circuit in the commutator.

1. Set the meter to the R×1 scale.
2. Touch the meter probes to any two adjacent bars on the commutator and note the meter reading.
3. Touch the meter probes to the next two adjacent bars on the commutator and note the meter reading.
4. Continue testing adjacent pairs of bars on the commutator and note the readings. If any one reading is substantially lower than the other readings, there is a shorted coil between the two bars with the low reading. If there is an unusually high reading between any two bars, there is a broken coil in the armature between the two bars.

# CHAPTER 7 • General Information

## AC AND DC

A bolt of lightning is direct current. (It is so direct that it can strip the bark off a tree!) Huge generators powered by water, wind, fossil fuels, or nuclear fission create direct currents of electricity that flow steadily off in only one direction. The generators push high-voltage electricity out across the country through mammoth cables.

DC has several drawbacks. Most significantly, the farther you send it, the more oomph it loses (this is called voltage drop), so there must be booster stations along the way to replenish the lost voltage. Electrical transmission over long distances was made much more efficient by the invention of alternating current.

AC is regularly fluctuating current that goes in one direction for a specific time, then reverses itself for the same time. In the United States AC makes this reversal 60 times per second, and therefore it is called 60-cycle AC, or 60-Hertz AC. If it fluctuated only once per second your lights would flicker visibly, but at 60 times per second the human eye cannot discern the change.

### *The Magic of Transformers*

The high-voltage electricity coming out of the giant generators must eventually be reduced by transformers before it can be used. DC enters one end of a transformer and zips through coils of successively smaller-gauge wires that reduce the voltage. By the time electricity enters your house, it is down to 120 volts. Voltage reduction also occurs in your car, which uses DC from the battery to start the gasoline engine.

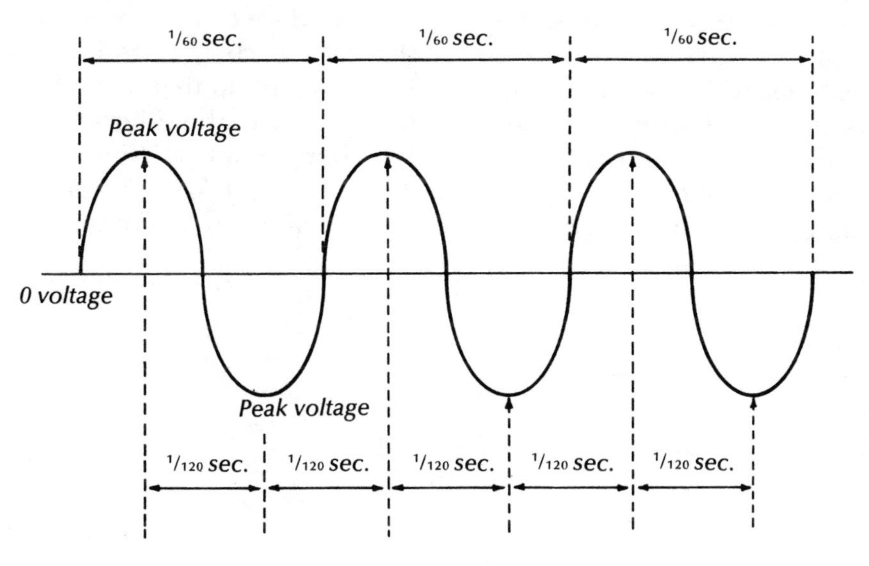


60-Hertz alternating current

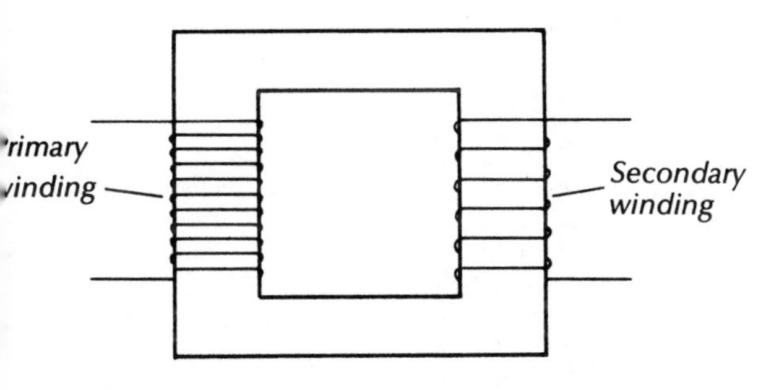


The basic arrangement of coiled wire inside a transformer. Current enters the primary winding and leaves the secondary winding at a lower voltage

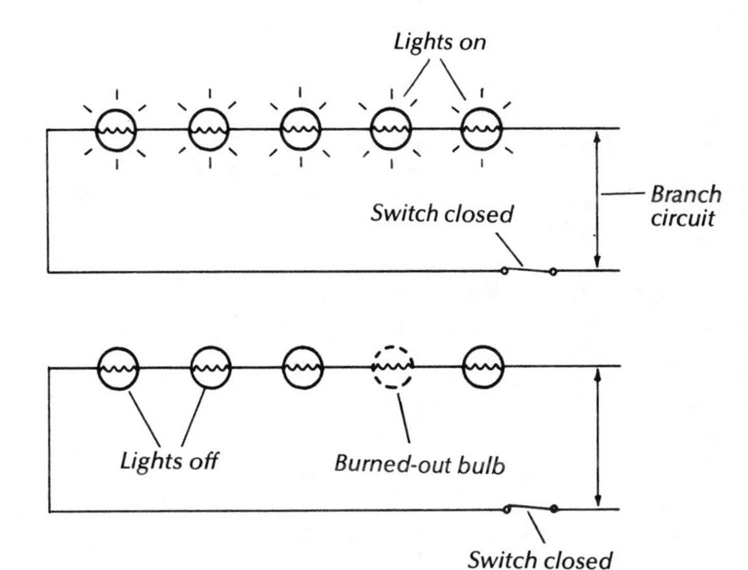


Series circuit. When one bulb burns out, none of the others will light

# SERIES AND PARALLEL CIRCUITS

Appliances, lamps, and switches are connected to the wires that feed them current in one of two ways. When devices are wired in series, interrupting power to one device on the circuit will affect all the devices. For example, if several switches are wired on one series circuit, turning off one switch will break the circuit, and none of the switches will receive power. With some Christmas tree lights, when one bulb burns out, none of the bulbs will light. Fuses and circuit breakers are always wired in series with the circuits they control, so that when something happens anywhere in the circuit, the interrupter will shut off all power to the circuit.

If the devices are connected in parallel, the branch circuit has been divided into mini-circuits, so that if something happens to one device, the others will still function. For example, if two lamps are plugged into an outlet, each can be turned on and off without affecting the other.

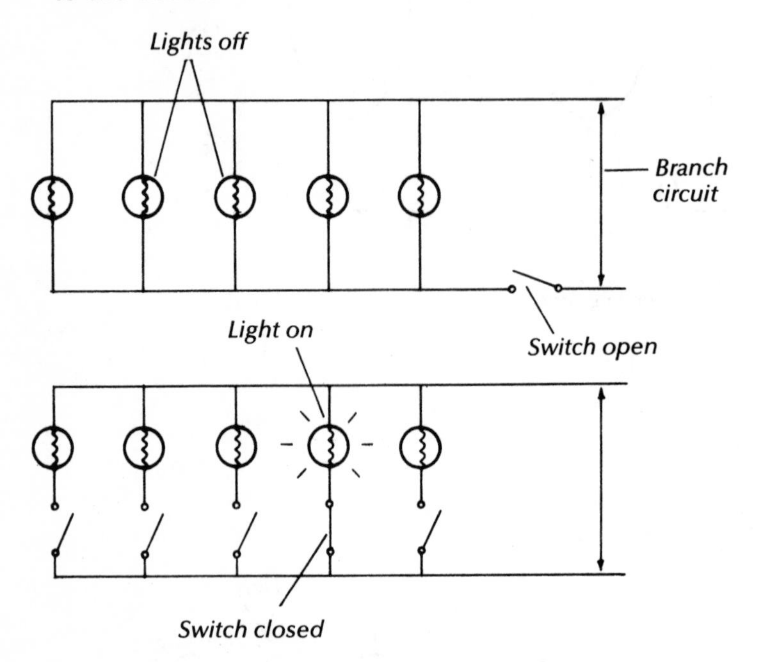


*Parallel circuit. Appliances wired in parallel can be controlled by a single switch (top) or separate switches (bottom). Turning off one or more does not affect the current going to the other units*

# WIRING SYMBOLS AND ABBREVIATIONS

It is space-consuming and difficult to draw a picture for every outlet and switch in an architectural electrical plan, so over the years symbols and abbreviations have been developed, which are used to diagram both house and appliance wiring. The symbols are often given an abbreviation to fully clarify what a part is.

| | |
|---|---|
| DT | Dust-tight |
| DW | Dishwasher |
| EP | Explosion-proof |
| G | Grounded |
| PS | Pull switch |
| R | Recessed |
| RT | Rain-tight |
| UNG | Ungrounded |
| WP | Weatherproof |

*Standard symbols used in wiring diagrams*

| *Ceiling* | *Wall* | |
|---|---|---|
| | | *Surface fixture* |
| R | R | *Recessed fixture* |
| X | X | *Surface exit light* |
| XR | XR | *Recessed exit light* |
| | B | *Blanked outlet* |
| J | J | *Junction box* |
| L | L | *Outlet controlled by low-voltage switch* |

| *Ceiling* | *Wall* | |
|---|---|---|
| | | *Fluorescent fixture* |
| R | R | *Recessed fluorescent fixture* |
| | | *Continuous-row fluorescent fixture* |
| R | | *Recessed continuous-row fluorescent fixture* |
| | | *Bare-lamp fluorescent strip* |

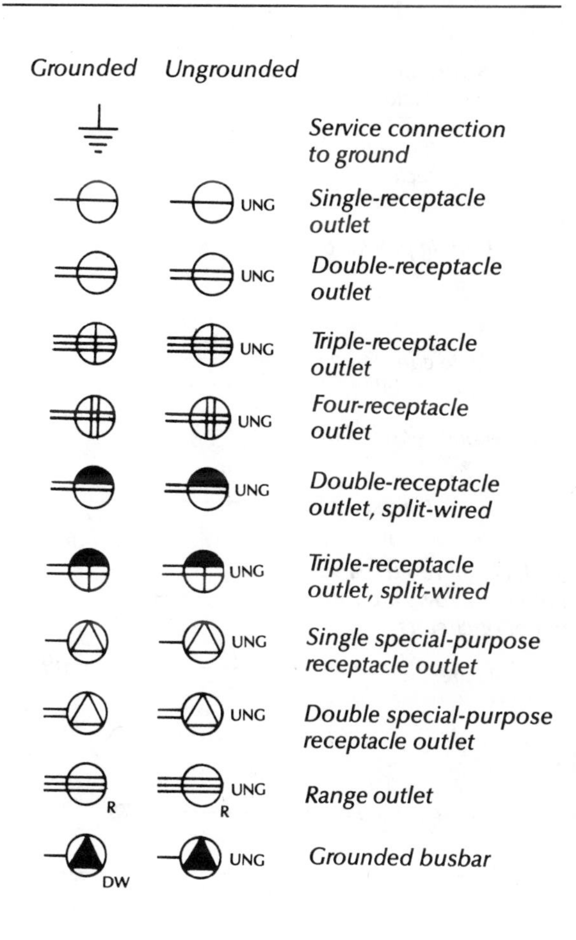

Grounded Ungrounded

Multiple-outlet assembly UNG

Clock hanger receptacle C C UNG

Fan hanger receptacle F F UNG

Switch and single receptacle S

Switch and double receptacle S

Ceiling pull switch S

Wire concealed in wall or ceiling

Wire concealed in floor

Exposed wire

Branch circuit run to panel board; number of arrows indicates number of circuits

3 wires (left), 4 wires

Wiring turned up

Wiring turned down

Push button

Buzzer

Bell

Combination bell and buzzer

CH Chime

Annunciator

D Door opener

M Maid's signal plug

Connection box

BT Bell-ringing transformer

Outside telephone

Interconnecting telephone

R Radio outlet

TV Television outlet

## INDEX

*Page numbers in italics indicate material in illustrations*


Adapter plug, *29,* 55–56
Add-on panel box, *8, 57, 71,* 71–72
Allen screws and wrenches, 91
Ampere/amperage, 43
  formulas, 44
Appliance cycles, 100–102
Appliance parts stores, 87, 99
Appliance surge, allowance for, 48
Appliances, 12, 105
  basic parts, 95–111
  cost to run, 48
  double-insulated, 56, 78
  grounded, 55–57, 78
  maintaining, 88–92
  repairing, 86–111
  short circuit in, 11, 56, 96
Appliances, high-voltage:
  separate circuit for, 48
  separate power disconnect for, 9
Appliances, large, 64
  components, 87–88
  disassembling, 88–92
    diagramming wires and parts, 91
  heating elements, 103
  reassembling, 92
Appliances, sealed, 90
Appliances, small:
  heating elements, 103
  obtaining parts for, 88
  repairing, 88
Armature (motor), testing for leakage, 111
Armored (BX) cable, 54–55, *82*

Branch circuits, 7
  adding, 49–74
  connecting to distribution panel, 58, 70–71
  general-purpose, 47
  major-appliance, 48
  overloaded, 48
  planning, 57–58
  running, 59–62
  service capacity, 47–48
  special-purpose, 47–48
Building codes, 50, 57–58, 76, 78, 79, 81, 85
Building-department inspection, 58, 63

Cable(s), 24, *53,* 56–57
  armored (BX), 54–55, *82*
  connecting to distribution box, 70, 85
  four-wire, 54, 65
  lead-encased, 76
  measuring and buying, 58
  nonmetallic sheathed, 54
  outdoor (weatherproof), 75, 76, 78
    burying, 76, 82
    estimating needs, 79
  running, 60–62
  securing to ceiling joists, 70
  service entrance, 76
  three-wire, 54, 64–70
  too short, 62
  two-wire, use in branch circuits, 64-70
  types, 54–55, 76
Cable clamp, 51, 80, 81
  types, 63–64
Cable connectors, 63–64
Capacitors, 107–108
Cartridge fuses, 13
Caulking outdoor wiring, 78–79, 82, 85
Circuit(s), 12
  appliance, 58
  capacity, 48
  general-purpose, 58
  grounded, 55–57
  maximum length, 58
  running new, 49–74
  open, testing for, 110–11
  outdoor:
    connecting to inside branch circuit, 82–83
    installing, 78–85
  parallel, 113–14
  series, 113–14
  three-wire, 45–46, 48
  two-wire, 45,
  wattage, *58*
  *See also* Branch circuits
Circuit breaker panel, 10, 14, 57
Circuit breakers, *8,* 13–14, 114
  amperage, 48
  convenience of, 12
  replacement by GFCI, 56
  resetting, 10, 14
  *See also* Circuit interrupters
Circuit check, 14–15
Circuit interrupters, 7–15
  amperage, 43, 47–48, 58
Clamps, cable, 51, 63–64, 80, 81
Color coding:
  cable, 54
  appliance timer, 99
  in large appliance, 91
  wires and screw terminals, 24–25, 26, 29
  wiring harness in appliance, 96
Commutator (Universal motor) testing, 111
Conduit, *53,* 76–78, 82, 84–85
  bending, 77
  estimating needs, 79

Connectors:
cable, 63–64
crimp-type, 21–22
*See also* Wire connection
Contacts, electrical, 91, 94, 106
Continuity test, 92, 94, *110*
*See also* Hot-line tester; Multimeter
Cord sets, appliance, 19–20, 86, 95–96, *96*
strain-relief devices, 95
testing, 95–96
wire gauge, 96
Cord wires, splicing, 19–20
Crimping tool, 21, 22
Crimp-on connectors, 21–22
Current:
AC, 108, 112, 113
DC, 108, 112
Current flow, testing for, 94
*See also* Continuity test

Dimmer switch, *28*
advantages of, 25
replacing, 27–28
Disassembling large appliances, 88–92
Dishwasher:
heating element, 103
timer chart, *101*
Distribution box/panel, 7, *8*, 45
connecting cable to, 70, 85
Door bells and intercoms, wire gauges used for, 52
*See also* House signaling systems
Double-insulated appliances, 56, 78

Electric meter, *see* Kilowatt meter
Electrical boxes, 24, 51, *52*
in ceilings, 60
connecting, 62–70, 81
ganging, 75
in lath and plaster, 59, 60
outdoor (weatherproof), 75–76, 81
position of, 80
stabilizing free-standing, 84
tabbed metal hangers in, 59, *60* in wallboard, 59

Electrical power, house:
shutting off, 9–10
turning on, with circuit breaker, 10
Electrical service entrance, 7, *8*, 54
Electrical system, house:
grounded, 56–57
preparing to upgrade, 43–48
repairs, 24–42
short circuit, 11, 14
wire connections, 16–17
Electrician's knot, *39*, 95

Electricity, 7, 44–47
Ell fittings in conduit, 77, 82, *83*
installing, 84–85
Extension cords, 79
splicing, 19–20

Fiber bushing, 55, 63
Field coil (motor):
testing for leakage, 111
testing for open circuit, 110–11
Fishing cable, 80
around doors and windows, 61–62
through conduit, 84
through lath and plaster, 61
Fish tapes, 60, *61*, *77*, *79*, 82
Fixture, installing on outside wall, 79–82
Fluorescent fixtures, 39–42
Four-wire cable, 54, 65
Fuse box, 10
Fuse puller, 13
Fuses, *8*, *45*, 57, 114
amperage rating, 11–13, 48
blown, causes, 10–11
capacity, 10–11
cartridge, 13
changing, 9, 13
testing, 13, 51
changing/replacing, 9, 11–13
nontamp (type-S), *10*, 11–12
plug, 11–12
screw-in breaker (resettable), 10, 12
standard, 10, 12
time-delay (Fusetron), 12
types, 11–12, 13
wire gauges for, 58
*See also* Circuit Interrupters

Gauge, *see* Wire gauge
General-purpose circuits, 47
Generators, 112
GFCI, GFI, *see* Ground-fault circuit interrupter
Grounded appliances, 55–57, 78
Ground-fault circuit interrupter, 56, 78, 82
Grounding, 19, 29–30, 56–57, 65
busbar, 14, 70, *71*
plug, three-prong, 55–56
adapter for, *29*, 55–56
wire(s), *25*
in cable, 55, 57, 63–64, 81
in combination outlet/switch, 33
in outlet, 30–32
in switch, 27

Hacksaw, use in stripping armored cable, 50, 54, 63, 82
Heating elements, appliance, 103–104

Hickey, 77, 82
High-voltage appliances, 9, 48
Hot-line tester, 13, 50–51, 70, 73
*See also* Continuity test
Hot lines, 45–46, 56, 64–65
House signaling systems, 52, 72–74

Incandescent lamps, *see* Lamps, incandescent
Insulation, stripping, *see* Stripping insulation

Jumper wire(s), 95–96
in combination switch/outlet, *32*, 33
in outlets, 30
Junction boxes, 62, 66–67, 72, 82
with conduit, 85
outdoor, 78–79

Kilowatt-hour, 46
Kilowatt meter, 7, *8*, 46–47
Knife, for stripping wire, 15–16

Lamp post, outdoor, 82, 83–84
Lamps, fluorescent, 39–42
Lamps, incandescent:
components, 37
installing new cord, 38–39
repairing, 37–39
replacing plug, *39*
sockets, 37–39
switches, 37
Light bulbs, weatherproof, 78
Light fixtures, *25*
ceiling:
concealing fixture box, 37
controlled by wall switch, 35, *36*
diagramming wiring before removing, 35
installing chain fixture, *36*
replacing, 33–37
types of hangers, *34*
circuit for, 58
outdoor (weatherproof), 76
*See also* Fluorescent fixtures

Magnetic field, 106
Main power disconnect, 7–10, 45, 47
*See also* Circuit breakers; Fuses
Main power line, 13
Main power panel, 7–10, 45, 57
Major-appliance circuits, 48
Master switch, *see* Main power disconnect
Mercury (silent) switch, 25
Motors, 106–11
appliance, 86, 87
capacitor, 106–108
components and operation, 106

Motors, (*continued*)
- lubricating, 106
- shaded-pole, 106, *107*
- split-phase, 106, *107*
- testing for resistance, 109
- troubleshooting, 109–11
- types, 106–11
- Universal, 108–11

Multimeter, 91–94, 96, 103–104
- *See also* Continuity test

Multipurpose tool, *15*, 21, *50*, 54

National Electric Code (NEC), 49
- fuses, 11
- outlets, 29
- soldering, 22
- splitting armored cable, 55
- voltage, 46

National Fire Protection Association, 49

Nonmetallic sheathed cable, 54

Ohms (resistance), 43, 44

Ohms-adjustment knob (multimeter), 93

Outdoor circuits, 78–85

Outdoor equipment:
- cable, 75, 76, 78
- light bulbs, 78
- outlets, 78, 81
- switches, 78

Outdoor wiring, 75–85

Outlets (receptacles), *25*, 29–32
- backwired, *30*, 31–32
- circuits for, 58
- diagramming wires in, 30, 31
- grounding, 29–30, 51
- maximum number on a circuit, 58
- multiple, 30
- outdoor, 78, 81
- replacing, 29, 30–32
- safety, *30*
- spacing of, 58
- testing, 51

Overload, 11, 15, 25

Owner's manual, appliance, 87, 88, 90, 100, 103

Parallel circuit, 113–14

Parts, appliance, 87, 99

Pigtail wire on adapter plug, *55*, 56

Power companies, local, 45

Power-generating plants, 44

Power lines from utility to house, capacity of, 57

Power surge, 12

Power-takeoff terminals, *57*, 71

Pump, appliance, 86, 87

Receptacle, *see* Outlets

Resettable fuse, 10, 12

Resistance, 52, 94
- formula, 48
- *See also* Ohms

Series circuit, 113–114

Service manual, appliance, 87, 88, 103
- typical exploded drawing in, *89*

Short circuit:
- in appliance, 11, 56, 96
- cause of blown fuse, 11
- cause of power interruption, 14
- causes of, 11, 77–78

Soldering, 22–23, 39, 78, 85

Solenoids, 104–105

Solid-state controls, 105–106

Special-purpose circuits, 47–48

Splicing wires, 18–20, 22–23, 62, 73–74
- cord wires, 19–20
- solid wires, 18–19
- stranded wires, 19

Strain relief:
- device, 95
- electrician's knot, *39*, 95

Stripping insulation:
- from cable, 54–55
- stripping guide on backwired switches, 28, *29*
- from wire, 15–16

Surge, allowance for, 48

Switches, 24–29
- appliance, 86, 97–98
- backwired, 28–29
- combination, 32–33
- diagramming wires, 26
- dimmer, 25–28
- hot wires in, 27
- installing/replacing, 25–29, 33
- outdoor, 78
- push-button and tap, 25
- silent (mercury), 25
- standard, 26–27
- types, 25, *26*

Tapping wires, 20–23

Template, 59

Terminals, appliance, 96–97

Thermostat, appliance, 86, 102–103
- bimetal strip, 102, 103
- gas-filled, 102, 103
- replacing, 87, 103
- testing, 103
- thermodisk, 102, 103
- types, 102, 103

Three-prong plugs, 55–56

Three-wire cable, 54, 64–70

Three-wire circuits, 45–46, 48

Timer, appliance, 86, 98–102
- motor, 98–99
- replacing, 87, 99–102
- testing, 94

Tools:
- for adding branch circuits, 49–51
- for installing conduit, 77

Transformers, 42, 112–13
- in house signaling system, 72–74

Trench:
- for outdoor cable/conduit, 83–85
- running across studs, *61*

Two-wire cable, 64–70

Two-wire circuits, 45

U-nails, 73

Underwriters' Laboratories, Inc., 49

Universal motor:
- brushes, 108–10
- components, 108, *109*
- special tests for, 111

Volt/voltage, 43, 46
- formula, 44

Voltage drop, 112

Washing machine:
- bad timer, 87
- power requirements, 45
- *See also* all entries for Appliances

Watt/wattage, 43–44
- circuit, *58*
- formula, 44, 48

Watt-hour, 46

Weatherproof equipment, *see* Outdoor equipment

Wire(s), wiring, 11, 52
- aluminum, 52
- copper, 52
- outdoor, 75–85
  - rules for, 77–78
- solid:
  - splicing, 18–19
  - tapping, 20
- stranded, 16, 52
  - splicing, 19
  - tapping, 20–21
- underground, 82–85
- *See also* Cable(s); Wire connection; Wire gauge

Wire connection, 15–23
- screw terminals, 16–17
- soldering, 22–23
- splicing, 18–20, 22–23, 73, 74
- stripping insulation for, 15–16
- types, 16–21
- wire nuts, 17–18, 36, 37, 51, 78, 85

Wire cutting tools, 54

Wire gauge, 33, 52, *53*, 54
  in appliance cord sets, 95
  in general-purpose branch circuits, 47
  in house signaling system, 72
  for major-appliance circuits, 58
  in motors, 106
  in special-purpose branch circuits, 47

Wire Gauge System, American, 52

Wire nuts, 51
  types, 17–18
  use in installing ceiling lighting fixture, 36, 37
  use in outdoor wiring, 78, 85

Wire stripper, *15*, *50*, 54, 82

Wiring arrangements:
  adding fixture or outlet to junction box, 66–67
  adding outlet to existing outlet, 66
  adding switch and outlet to existing fixture, 66
  adding switch and outlet in same box to fixture, 67
  controlling a fixture between two switches, 67–68
  controlling one fixture with three-way and four-way switches, 69–70
  fixture controlled by three-way switches; outlet connected to fixture is always hot, 68–69
  for house signaling system, *72*

Wiring arrangements, (*continued*)
  one fixture controlled by two different switches, 67
  separate fixtures controlled by different switches in same box, 68
  two fixtures controlled by separate switches, 65–66
  wall switch controlling one light, 65

Wiring harness, appliance, 86
  color-coded, 96
  following during continuity test, 94

Wiring list, 48

Wiring symbols and abbreviations, 115–16